Nontariff Barriers

The Effects on Corporate Strategy in High-Technology Sectors

About the Book and Authors

This book explores the impact of Japanese and European nontariff barriers (NTBs) on the international marketing, investment, and technology strategies of small- to medium-sized high-technology U.S. firms. The study documents a pronounced dichotomy between strategies of small and large companies that, to a significant extent, reflects the gap in the resources, bargaining power, and familiarity with foreign markets of these two groups. Conclusions concerning the efficacy of corporate strategies adopted are supported by an analysis of over 20 case studies. These strategies range from licensing agreements with local firms to use of a trading company or local distributor, formation of a joint venture with a local firm, and establishment of a wholly owned subsidiary in the foreign market.

Michael F. Oppenheimer is executive vice president at The Futures Group, where *Donna M. Tuths* is senior research associate in the International and Technology Section.

Nontariff Barriers

The Effects on Corporate Strategy in High-Technology Sectors

Michael F. Oppenheimer
and Donna M. Tuths

Westview Press / Boulder and London

Westview Special Studies in International Economics and Business

This Westview softcover edition is printed on acid-free paper and bound in softcovers that carry the highest rating of the National Association of State Textbook Administrators, in consultation with the Association of American Publishers and the Book Manufacturers' Institute.

Copyright © 1987 by the Futures Group

Published in 1987 in the United States of America by Westview Press, Inc.; Frederick A. Praeger, Publisher; 5500 Central Avenue, Boulder, Colorado 80301

Library of Congress Catalog Card Number: 87-050073
ISBN: 0-8133-7403-0

Composition for this book was provided by the authors.
This book was produced without formal editing by the publisher.

Printed and bound in the United States of America

(∞) The paper used in this publication meets the requirements of the American National Standard for Permanence of Paper for Printed Library Materials Z39.48-1984.

6   5   4   3   2   1

CONTENTS

PREFACE

The research for this book was conducted for the Office of Advocacy of the Small Business Administration in Washington, D.C., over an eleven-month period ending in September 1985. The basic objective of this study was to explore the impact of Japanese and European nontariff barriers (NTBs) on the international marketing efforts and investment and technology strategies of small- to medium-sized high-technology U.S. firms. Underlying this effort was the belief that NTBs have a greater impact on smaller firms than on larger ones because of the less-extensive international presence of most small businesses, their lack of sophistication in international trade matters, and the lack of resources such companies can devote to responding to NTBs.

Our previous work in this area convinced us that the most useful NTB research in support of trade policymakers is to provide detailed case histories describing the interaction of high-technology firms with the import regimes of our principal trading partners. The status of NTB data was reviewed in some detail in our previous work, wherein we found several NTB data bases of significant volume but lacking in timeliness, consistency, and detail.[1] We were particularly determined to go beyond a simple inventory of NTBs to describe their workings at the company level, and to improve our understanding of the strategies that small- to medium-sized firms adopt in the presence of barriers to market entry. Thus, it is the interactions among market distortions, corporate strategy, and U.S. trade policy that occupy us in this report.

For purposes of this study, small- to medium-sized firms were identified by the Office of Advocacy as those companies whose total assets do not exceed $50 million. To determine the differential impact of these barriers, both large and small high-technology firms were interviewed so that the unique behavior of small firms would be highlighted. Over 25 percent of the case studies included in this report fall into the small- to medium-sized category defined by the SBA as appropriate for these high-technology sectors.

The process of identifying and eliciting case material from these firms proved a greater task than had been anticipated. A large number of the firms contacted were unwilling to participate in the study because they did not recognize nontariff barriers as such--but rather considered the difficulties they encountered as legitimate aspects of doing business in an affected market--or because they recognized the existence of market distortions and their effect on firm behavior, but did not want to talk to the study team for fear of foreign government retaliation or of leaking important competitive information to other firms. As a result, case material was difficult to obtain.

In the process of this search, the study team conducted over seventy face-to-face interviews with individuals from industry, government, consulting, and various trade organizations. Several hundred leads produced by these efforts were followed up, resulting in the twenty-three cases discussed in this book. The companies interviewed encountered a wide range of market access problems. Among those, four main nontariff barrier types emerged--standards, performance requirements, government procurement, and patents. Discussion of these NTBs can be found at the end of each industry chapter immediately preceding the case material.

Not surprisingly, an analysis of the strategies adopted by small- to medium-sized firms and large firms in response to these market barriers revealed a pronounced dichotomy that, to a large extent, reflects the gap in the resources of these two groups. Large companies are more familiar with the intricacies of foreign markets and will not, in general, go to the U.S. government for assistance in penetrating a foreign market. These firms tend to have greater bargaining power in the foreign marketplace because of the large number of jobs and the size of the investment they can generate. Thus, large firms have a wider range of options available to counter market barriers. Smaller firms, on the other hand, generally do not have substantial export experience nor large sources of capital to invest in risky ventures. As a result, their options are greatly limited. By obtaining actual case study material, the study team gained useful insight into the different strategies adopted by these two groups.

Conclusions concerning the efficacy of corporate strategies adopted were apparent from an analysis of the case studies. These strategies ranged from licensing agreements with local firms, to use of a trading company or local distributor, formation of a joint venture with a local firm, and establishment of a wholly owned subsidiary in the foreign market. The case material revealed that some strategies were more prone to complications than others, regardless of the size of the firm implementing a given strategy. Overall, the licensing agreements encountered in this study did not prove to be effective over the long run. Companies found that the licensing arrangement was too difficult to control from a U.S. base. Consequently, the arrangement developed in an unwieldy manner and led to the licensee's development of a competitive product which it would then sell directly into its domestic market.

On the whole, joint-venture agreements with local firms provided an effective means of penetrating a market blocked by preferential government procurement, standards problems, or significant performance requirements. These arrangements were found to be more stable over time if provisions for the transfer of technology and rights to the use of that technology by the joint-venture partners in world markets were narrowly defined in legal terms, and if the local partner was not a potential competitor of the U.S. firm.

However, the most effective means of circumventing market restrictions was found to be the establishment of a wholly owned subsidiary in the overseas market. This strategy allowed the U.S. firm to develop a critical understanding of the target market and local regulatory processes through the hiring of local nationals and proximity to the market. These firms were able to successfully confront standards problems and satisfy performance requirements with greater ease and fewer compromises to overall corporate goals. Though for the most part this strategy was adopted by mature companies with considerable financial resources, two case studies indicated that smaller firms should consider this option if the potential market is large enough and risk-taking can be adequately managed.

Michael F. Oppenheimer
Donna M. Tuths

NOTES

1. See Robert Cohen, Richard Ferguson, and Michael Oppenheimer, Nontariff Barriers to High-Technology Trade (Boulder: Westview Press, 1985).

ACKNOWLEDGMENTS

The authors would like to thank the numerous members of the U.S. business community who generously donated their time to this study and shared with us their valuable experiences in the West European and Japanese marketplaces. Without the contribution of these individuals, who at their own request will remain anonymous, this study would not have been possible. Special thanks are also given to Thomas Gray and Richard Boden of the Office of Advocacy of the Small Business Administration, which provided the funding for this research.

The authors would also like to acknowledge the superb editorial assistance provided by Beverly Pitts and the industrious efforts of Gail Layden in preparing the final manuscript for publication. Cathy Johnson, June Osborne, and Marion Healy also deserve an additional note for the long hours spent deciphering interview tapes.

Though the research for this study was originally supported by the Office of Advocacy at the Small Business Administration, the SBA bears no responsibility for the findings, analyses, and conclusions contained in this volume.

M.F.O.
D.M.T.

1
EXECUTIVE SUMMARY

This study was conducted for the Office of Advocacy of the Small Business Administration in Washington, D.C., over an eleven-month period ending in September 1985. The basic objective of this study was to explore the impact of Japanese and European nontariff barriers (NTBs) on the international marketing efforts and investment and technology strategies of small- to medium-sized high-technology U.S. firms. Underlying this effort was the belief that NTBs have a greater impact on smaller firms than on larger ones because of the less-extensive international presence of most small businesses, their lack of sophistication in international trade matters, and the lack of resources such companies can devote to responding to NTBs.

Our previous work in this area convinced us that the most useful NTB research in support of trade policymakers is to provide detailed case histories describing the interaction of high-technology firms with the import regimes of our principal trading partners. The status of NTB data was reviewed in some detail in our previous work, wherein we found several NTB data bases of significant volume but lacking in timeliness, consistency, and detail. We were particularly determined to go beyond a simple inventory of NTBs to describe their workings at the company level, and to improve our understanding of the strategies that small- to medium-sized firms adopt in the presence of barriers to market entry. Most observers are accustomed to describing the effects of NTBs as "subtle," but what precisely are these subtle effects? How do firms--and small- to medium-sized high-technology firms, in particular--alter their marketing, invest-ment, technology licensing, and R&D strategies in the presence of these barriers? How do the firms themselves weigh the costs and benefits of these strategies over time? Do the subtle adjustments that firms make in response to NTBs raise policy issues not

1

adequately addressed currently? What practical problems exist in framing a trade negotiating agenda around barriers with which U.S. firms--in making their strategic adjustments--implicitly comply? In short, it is the interactions among market distortions, corporate strategy, and U.S. trade policy that occupy us in this report.

For purposes of this study, small- to medium-sized firms were identified by the Office of Advocacy as those companies whose total assets do not exceed $50 million. To determine the differential impact of these barriers, both large and small high-technology firms were interviewed so that the unique behavior of small firms would be highlighted. An examination of Table 1.1 shows that over 25 percent of the case studies included in this report fall into the small- to medium-sized category defined by the SBA as appropriate for these high-technology sectors.

The process of identifying and eliciting case material from these firms proved a greater task than had been anticipated. A large number of the firms contacted were unwilling to participate in the study because they did not recognize nontariff barriers as such--but rather considered the difficulties they encountered as legitimate aspects of doing business in an affected market--or because they recognized the existence of market distortions and their effect on firm behavior, but did not want to talk to the study team for fear of foreign government retaliation or of leaking important competitive information to other firms. As a result, case material was difficult to obtain. In the process of this search, the study team conducted over seventy face-to-face interviews with individuals from industry, government, consulting, and various trade organizations. Regional and state SBA offices were contacted by mail and phone to explain our study objectives and request potential sources of case study material. Several hundred leads produced by these efforts were followed up, resulting in the twenty-three cases summarized in Table 1.1. The companies interviewed for this study encountered a wide range of market access problems. Among those, four main nontariff barrier types emerged--standards, performance requirements, government procurement, and patents. Table 1.2 lists the types of NTBs most commonly encountered within each industry sector and provides examples of case studies in which these barriers were encountered; further discussion of these NTBs can be found at the end of each industry chapter immediately preceding the case material.

Not surprisingly, an analysis of the strategies adopted by small- to medium-sized firms and large firms in response to these market barriers revealed a pronounced dichotomy that, to a large extent, reflects the gap in the resources of these two groups.

TABLE 1.1
Summary of case studies

Firm No.	Total Assets ($ Mil)	Strategy	Reason
1	55	Joint venture in Japan with nonpharmaceutical company; limited technology transfer	Need for non-U.S. source of manufacture to circumvent U.S. regulations. Need for local partner to deal with Japanese regulatory authorities and system need for capital; desire to retain rights to U.S. market
2	44	Sells through Japanese trading company; technology transfer outside of company's area of proprietary research; in Europe has own marketing representative	Perceived difficulty of marketing in Japan due to cultural differences and complex regulatory structure; need for capital, but desire to maintain an area of proprietary research
3	over 500	Wholly owned marketing and service organizations throughout Europe with a manufacturing facility in Germany; two joint ventures with local Japanese companies; greater emphasis on developing products specifically for the Japanese market in the future	Nationalistic attitudes throughout Europe; "buy national" in France; difficulty selling to PTTs in Europe; teamed with local Japanese companies that were not competitors; heavy competition in large-scale computers from local Japanese companies and appearance of regulation in that market
4	502	In Europe and Japan, began with distributors and then set up wholly owned subsidiaries for sales and marketing; also wholly owned manufacturing subsidiaries in Europe	Increasing pressure to contribute to local economy; pressures to perform local R&D, joint R&D with local firm, or source from local companies; difficulty retrieving some of the costs associated with adapting the product to Japanese market
5	88	Licensed product in Japan and lost market there	Need for capital; did not do sufficient R&D to create products for local Japanese market; unable to obtain patent in Japan
6	over 100	Joint venture in Austria for manufacturing and design; Design Center in Japan	Circumvent duty system; enable interactive custom design; difficulties selling to European PTTs (specifications used to favor local vendors); broadly defined NTT specifications; nationalistic attitudes
7	96	Joint venture in Japan since the 1950s for sales and marketing only (not very successful); hope for wholly owned manufacturing in Asia in the future; no great optimism for European market	Unable to obtain patents in Japan; problems with "slippery" standards in Japan and Europe; nationalistic attitudes in Europe and preference for Siemens in Germany

(Continued)

TABLE 1.1 (Cont.)

Firm No.	Total Assets ($ Mil)	Strategy	Reason
8	265	Wholly owned sales and manufacturing subsidiaries employing local nationals	Strong commitment to "making it on its own" and desire to maintain control
9	25	Joint venture in Japan for sales, manufacturing, R&D for local market; wholly owned subsidiaries in Europe	Problems with distributor relationships in Europe (copied product) and Japan (no contact with customer); difficulty obtaining financing for 100 percent owned facility in Japan; received pressure from French government to form a joint venture with local firm and transfer technology
10	65	Began with trading company in Japan and later established joint venture for sales and manufacturing; wholly owned sales subsidiaries in Europe with manufacturing site in France	Need to have more direct involvement in Japan because of market size and importance; difficulty hiring salesmen and engineers in Japan; unable to participate in MITI-funded research
11	6	Licensing agreement in Europe that resulted in a breach of contract; licensing agreement in Japan that had to be terminated; currently using distributor in Japan	Loss of technology through licensing agreements in Europe and Japan
12	300	Originally used trading company in Japan for distribution and service; in 1980 established joint venture in Japan for service, with intent to develop sales and manufacturing there; in Europe sold through distributors until 1973, then established a sales and later manufacturing subsidiary in Europe; in 1981, established joint venture in France for R&D manufacturing and sales	MITI-funded development of competing technology given to Japanese firm and Japanese customers urged to buy from this local company; Japanese targeting of domestic industry; Japanese customers began to require customization of product that could not be provided from U.S. manufacturing site
13	190	Joint venture in Japan that resulted in loss of technology; after this experience, used a local distributor in Japan until recently, then a joint venture for sales and manufacturing was established	Loss of technology through a joint venture in Japan; as Japanese catch up technologically, company must provide more and more modification of product; inability to hire work force unless affiliated with large company in Japan

(Continued)

TABLE 1.1 (Cont.)

Firm No.	Total Assets ($ Mil)	Strategy	Reason
14	1	Delayed entry to Japanese market	Inability to obtain adequate patent protection in Japan
15	200-300	Joint venture with local partner in two European countries and Japan; joint R&D in these countries with R&D also at headquarters	Need to look like a local company; need for understanding of local client; EC procurement policies obscured information for nonlocals
16	150	Distributors around the globe with wholly owned sales subsidiaries in key countries; wholly owned manufacturing and R&D facility in Europe; joint R&D in Japan	Need to show commitment to the British market to sell to nationalized buyers; need to work with a local Japanese company to properly transform product
17	10-15	In Europe, began with distributors and slowly replaced with joint ventures for sales and service in Japan, work with trading company	Need to work closer with customer; difficulty selling to NTT; no appreciable amount of business with European PTTs; difficulty selling to Japan because of Israeli manufacturing site
18	65	In Japan, works with value-added reseller to which it provides all sensitive supporting documentation such as specifications and source code in exchange for worldwide rights (excluding Japan) to market any enhancements to the base technology developed by the Japanese reseller	Local firm helps in certification of company's product in Japan; perception of improved service capabilities in local market; commitment to sell to NTT
19	85	Originally had distributor in Europe, but terminated the arrangement after it learned that distributor had copied technology; later wholly owned sales and service subsidiaries and manufacturing facilities in Europe	European distributor copied technology and sold its own competing product; needed in-country presence to be considered a reliable supplier; service is key; "buy national" in France; pressure to manufacture locally in France; difficulty with German PTT and qualification testing; incompatible CCITT standard in Europe; difficulty with NTT standards
20	105	Uses own U.S.-based sales force to sell overseas	Difficulties due to CCITT standards used in Europe; preference for local companies shown in government procurement; need to bid through local

(Continued)

TABLE 1.1 (Cont.)

Firm No.	Total Assets ($ Mil)	Strategy	Reason
20 (Cont.)			company in Germany; perceives Japanese market as completely closed and does not pursue
21	63	Sells through an international distributor; in Japan, selected an individual to act as agent	In France, difficulty selling to PTT, which became more severe once its local distributor was nationalized; in Germany, received pressure from the Bundespost to transfer technology to several local firms; overall, nationalism a problem in Europe; certain limits to the distributor relationship when times are bad; in Japan, perceived the market to be closed during the 1970s and did not attempt to enter, in the 1980s adopted a cautious approach
22	under 50	Uses distributors	In Germany, Bundespost has used standards to urge company to work with local company, also problem with export of sensitive data; problem of incompatibility with CCITT standards in Europe
23	under 50	Joint ventures with local partners	Uses local partners to deal with local regulatory authorities; some problems with German Bundespost which uses standards to manipulate terms of doing business; "buy national" in the United Kingdom, Germany; opposition from the KDD was overcome by local Japanese partner

Large companies are more familiar with the intricacies of foreign markets and will not, in general, go to the U.S. government for assistance in penetrating a foreign market. These firms tend to have greater bargaining power in the foreign marketplace because of the large number of jobs and the size of the investment they can generate. Thus, large firms have a wider range of options available to counter market barriers. Smaller firms, on the other hand, generally do not have substantial export experience nor large sources of capital to invest in risky ventures. As a result, their options are greatly limited. An examination of the case

TABLE 1.2
Types of NTBs encountered by industry

Industry	NTB Type	Case Study Number
Biotechnology:	Performance requirements	2
Computers:	Government procurement	3
	Performance requirements	4
	Patents	5
Semiconductors:	Government procurement	6
		7
		12
	Standards	6
		7
	Patents	7
	Performance requirements	9
Software:	Government procurement	15
		16
	Patents	14
Telecommunications:	Standards	18
		19
		20
		22
		23
	Government procurement	17
		19
		20
		21
		23
	Performance requirements	19
		21
		22

studies provides useful insight into the different strategies adopted by these two groups.

The study found that small- to medium-sized firms confronting trade-restricting standards in an overseas market are likely to seek the representation of a domestic firm to deal effectively with local regulatory bodies. Within this group, the smaller and less-experienced firm, for example, opted for a relationship with a local distributor or trading company that is familiar with the local regulatory system and can ease the entry of the firm's product to the market. This strategy was adopted by the firm in Case No. 2, for example. This young biotechnology firm never seriously considered any other option because of the complex regulatory processes involved in the commercialization of biotechnology products. However, examples of this are more common in the telecommunications industry because of the frequent use of standards by the local PTTs to discriminate against nonlocal products and the complex certification processes throughout Western Europe and Japan. Thus, examples can be found in Cases No. 17, 18, 21, and 22.

Conversely, a larger firm, confronted by these same standards issues will use its considerable resources to negotiate a way around the barrier through its local sales subsidiary, or will establish a production facility within the market and manufacture to local specifications. For example, though the firm in Case No. 19 began with a distributor in Europe, it eventually set up its own wholly owned sales and service and manufacturing subsidiaries in Europe. Large firms have been more successful at dealing effectively with these standards problems because they have the resources to establish staff and operations in foreign markets employing local nationals who are familiar with the practices of the host nation, have contacts within the local regulatory system, and can react quickly to any problems that arise during the certification process. Unfortunately, the large capital expenditures associated with this strategy militate against its use by smaller firms.

The reaction of small and large firms to the imposition of performance requirements did not vary that greatly, though some difference can be noted. Both sets of firms are wary of licensing technology to local firms in response to government pressure. For example, the small company in Case No. 9 waited a year and a half to establish a wholly owned subsidiary in France because of pressure from the French government to form a joint venture with a local firm. Similarly, the large company in Case No. 4 has not to date been involved in any substantial joint R&D efforts in France despite persistent requests from the French government. In instances such as these, the government's intent to boost its local technological base is obvious, and therefore both groups of firms recognize the potential for significant technology loss. However, larger firms, because of the greater resources they command, have more options available to them in satisfying a local government's request. When possible, these firms will try to source or manufacture locally, rather than transfer any significant technological know-how. Moreover, the large firm tends to be in a better bargaining position vis-a-vis the local government because of its potential to generate significant local employment and export revenues. Thus, if the large firm does choose to work with a local firm, it is able to sufficiently narrow the legal definition of that arrangement to preserve its proprietary know-how and avoid the emergence of direct competition from this relationship.

The effects of preferential government procurement are demonstrated by Case No. 20, in which the closed nature of NTT procurement caused the firm to dismiss that potential market altogether. Case No. 21 is similar; for the first ten years of business the firm did not even try to sell into the Japanese market. This suggests a tendency among smaller firms to avoid a market that they perceive to be heavily influenced by "buy national" policies and to concentrate efforts elsewhere. On the other hand, a larger firm is more likely to manufacture within the country or otherwise contribute significantly to the local economy. In this way, the large firm makes it more difficult for the local government to discriminate against its product.

Finally, patent issues affect small and large firms differently. The small firm in Case No. 14 chose to delay its entry to the Japanese market until its patent dispute was settled. Because this firm's viability in the long run is significantly influenced by the success of this particular product in the Japanese market, the firm does not want to risk having its technology copied by the Japanese with no legal recourse. Large, more mature firms as a rule have more diversified product lines and are not dependent on a single technology for their success. Further, these firms have more resources to allocate to R&D projects and are constantly developing new products. As a result, larger firms confronted with the inability to obtain adequate patent protection in a market may enter that market despite the lack of protection and innovate as quickly as possible; Case No. 7 provides an example of this strategy.

Certain conclusions concerning the efficacy of corporate strategies adopted are apparent from an analysis of the case studies. As mentioned above, these strategies ranged from licensing agreements with local firms, to selling through a trading company or local distributor, forming a joint venture with a local firm, and setting up a wholly owned subsidiary in the foreign market. Analysis of the case material revealed that some strategies were more prone to complications than others, regardless of the size of the firm implementing a given strategy.

Overall, the licensing agreements encountered in this study did not prove to be effective over the long run. In a number of cases (Cases No. 9, 11, and 19), the licensing of technology to a local firm in a foreign market led to the loss of technology in both the European and the Japanese markets. Companies found that the licensing arrangement was too difficult to control from a U.S. base. Consequently, the arrangement developed in an unwieldy manner and led to the licensee's development of a competitive product which it would then sell directly into its domestic market. In most instances, because of the geographical distances involved and the licensor's lack of familiarity with the overseas market and customers, the licensee was able to develop this competing capability without the knowledge of the licensor. In the majority of these cases, the U.S. firm chose not to seek legal remedies because of the time and money involved in pursuing such a course in a foreign country. Instead, the licensing agreements were terminated and new strategies were developed. Case No. 2, however, provided an interesting example of one small- to medium-sized firm that was able to effectively manage licensing agreements. In this case, the firm has been successful because it is able to license technology in selected areas while maintaining certain niche product markets for itself. For small companies in a position to segment their R&D and product development in this fashion, this can be an effective means of circumventing market barriers, while increasing overall revenues.

Many companies retain local trading companies or distributors because of limited funds for the exploitation of overseas markets and unfamiliarity with foreign regulatory bodies, business practices, and consumer preferences. Nevertheless, over time, these relationships became less satisfactory because they insulated the U.S. company from its overseas customers and retarded the firm's ability to react to developments in the overseas market. In addition, important international marketing skills were not developed. Consequently, as the overseas market became more important to the firm, the distributor relationship became a less-effective means of reaching that market and the firm began to consider a direct sales presence. In Case No. 21, for example, the firm used an international distributor and found that its products tended to do well only in countries where the distributor did not have a competing product of its own. Thus, though the distributor relationship may be an effective means for entering a market, it is, in general, not a viable strategy over the long run.

On the whole, joint-venture agreements with local firms proved to be an effective means of penetrating a market blocked by preferential government procurement, standards problems, or significant performance requirements. These arrangements were found to be more stable over time, however, if provisions for the transfer of technology and rights to the use of that technology by the joint-venture partners in world markets were narrowly defined in legal terms, and if the local partner was not a potential competitor of the U.S. firm. In Case No. 13, the U.S. firm incurred a serious loss of technology as a result of a joint-venture arrangement in Japan that did not have sufficient legal safeguards. The Japanese joint-venture partner did have business interests that overlapped those of the U.S. firm prior to establishment of the joint venture. Other joint ventures have, however, fared rather well. Case No. 3, for example, describes two Japanese joint ventures in the computer industry. This U.S. company consciously chose to team up with Japanese firms that were not potential competitors and thus would view the arrangement as a profit-generating venture rather than a way to gain needed technological know-how. This company's decisions have yielded two Japanese joint-venture arrangements that are profitable and have remained stable for over twenty years. Several of the smaller companies interviewed have successfully implemented this strategy (Cases No. 9, 10, and 17). In Case No. 9, the firm's approach was to establish joint ventures overseas with local individuals with whom the firm had already established a trusting relationship through previous business transactions. Consequently, the firm was able to create autonomous overseas business ventures that benefit from local part-ownership. These ventures are profitable for the overseas partner and, therefore, there are no competing interests involved and there is no incentive to steal technology. Further, this type of joint venture, initially only for marketing and sales, requires little capital to start up. Overall, the study team found this strategy to be the most successful for small- to medium-sized firms.

The most effective means of circumventing market restrictions has, however, been the establishment of a wholly owned subsidiary in the overseas market. This strategy allows the U.S. firm to develop a critical understanding of the target market and local regulatory processes through the hiring of local nationals and proximity to the market. In this way, the firm is able to successfully confront standards problems and satisfy performance requirements with greater ease and fewer compromises to overall corporate goals. Moreover, as the firm's contribution to the local economy through employment and capital investment for facilities increases, the firm bypasses market-restricting preferential government procurement and, in fact, may even benefit from these policies. Though for the most part this strategy is adopted by mature companies that have considerable financial resources, Cases No. 9 and 10 indicate that smaller firms should consider this option if the potential market is large enough and risk-taking can be adequately managed.

2
INTRODUCTION

The basic purpose of this study has been to establish the impacts of Japanese and European nontariff barriers (NTBs) on the international marketing, investment, and technology strategies of small- to medium-sized high-technology U.S. firms. The context for the study is an increasing concern about NTBs as impediments to international marketing by U.S. companies, the contribution of these impediments--modest though it may be--to the huge imbalance in our international trade, and the growing presence of the Japanese in leading-edge industries, a presence viewed here in both international commercial and strategic terms. The ongoing bilateral discussions with the Japanese on high-technology access to Japanese markets, and U.S. efforts to place high-technology trade on the agenda for the upcoming multilateral trade negotiations, are manifestations of this increased concern. This report, therefore, represents an effort to establish a broader and more detailed empirical basis for approaching these high-technology trade issues, with particular focus on those small- to medium-sized companies that frequently define the leading edge in these industries.

Our previous work in this area[1] convinced us that the most useful NTB research in support of trade policymakers would be to provide detailed case histories describing the interaction of high-technology firms with the import regimes of our principal trading partners. The status of NTB data was reviewed in some detail in our previous work, wherein we found several NTB data bases of significant volume but lacking in timeliness, consistency, and detail. We were particularly determined to go beyond a mere inventory of NTBs to describe their workings at the company level, and to improve our understanding of the strategies that firms adopt in the presence of barriers to market entry. Most observers are accustomed to describing the effects of NTBs as

13

"subtle," but what precisely are these subtle effects? How do firms--and small- to medium-sized high-technology firms, in particular--alter their marketing, investment, technology licensing, and R&D strategies in the presence of these barriers? How do the firms themselves weigh the costs and benefits of these strategies over time? Do the subtle adjustments that firms make in response to NTBs raise policy issues not adequately addressed currently? What practical problems exist in framing a trade negotiating agenda around barriers with which U.S. firms--in making their strategic adjustments--implicitly comply? In short, it is the interactions among market distortions, corporate strategy, and U.S. trade policy that occupy us in this report.

Underlying this study is the increasing prominence of nontariff barriers as major commercial and diplomatic issues. A huge U.S. trade deficit and the perception that this deficit results in part from discrimination in major markets against U.S. products that would normally be competitive are part of the explanation for the growing controversy over NTBs. However, even if we found ourselves in a calmer trade policy environment, NTBs would dominate the trade negotiating agenda and would exist as important impediments to market access for U.S. firms. It is vital that these firm and industrywide effects not be lost in the debate over NTBs' contribution to the overall trade imbalance. Though this contribution is certainly modest, barriers to entry that deprive leading-edge firms of appropriate returns to R&D, and thus limit their longer-term innovative capacity, ought to be understood as major policy issues as we attempt to compete successfully in tomorrow's markets. It is important, therefore, that we understand the more secular, long-term forces contributing to the incidence and importance of NTBs and that we not expect them to diminish as more sensible currency alignments and shifting international growth patterns correct the current trade deficit.

The more enduring causes for the prominence currently accorded NTBs are complex, but fall generally into three broad categories: (1) those that are a consequence of either cyclical or secular market forces, some of which produce restrictions that are temporary and controllable through multilateral standstill agreements, bilateral negotiations, and the like, some of which are more permanent and may require more negotiating rounds and codes of conduct; (2) those NTBs that are policy-driven and reflect a tendency toward ever-broadening definitions of protectionism to encompass policy instruments previously understood as domestic; and (3) those restrictions that are a consequence of fundamental differences between major trading states in economic philosophy and organization, and a trend toward a more prominent government role in creating international comparative advantage in key high-technology sectors. These three alternate explanations for NTBs are not mutually exclusive (some may explain protectionism in mature sectors of industrialized country economies; others may explain protectionism in high-technology sectors), but they do tend to produce divergent explanations for the increase in protectionism overall. The perspective adopted, therefore, governs attitudes toward the feasibility and importance of trade negotiations in economic relationships among the major trading countries, the significance that should be attached to trade barriers in U.S./European and U.S./Japan relationships, as well as the modes and objectives for such negotiations.

Market-based perspectives on protectionism view NTBs as a natural and predictable reaction to cyclical periods of low economic growth, superimposed on longer-term structural changes in international competitiveness, the "trading-up" of newly industrializing countries, long-term price advantages stemming from disparate costs for factors of production, and the increasing ease of international technology transfer. Explanations for the current wave of protectionist pressure, particularly the United States, would also emphasize misalignment of currencies as well as asymmetries in growth patterns among the major industrialized trading countries. Policy responses here focus on standstill agreements by which the major trading countries agree to resist protectionist pressures and refrain from new trade barriers; engagement in continuous, formal negotiations (to maintain diplomatic and domestic political pressures against protectionism); domestic adjustment policies (to shift resources toward potentially more competitive sectors and ease the adjustment burden on trade-impacted industries and communities); and (prospectively) international agreements on adjustment designed to promote nonprotectionist responses to shifts in international competitiveness and contain protectionism within certain industries/modes of protection/time periods. It is assumed implicitly that protectionism is an aberration and irradicable through coordination of macroeconomic policies among the major trading countries, continuous aggressive international negotiating, and the international reinforcement of domestic political will in resisting protectionist pressures. The elimination of nontariff barriers is, therefore, viewed as a logical extension of previous negotiations on the reduction or elimination of tariffs, and part of a natural progression toward the elimination of all distortions to international trade.

A contrasting explanation for nontariff barriers emphasizes the dynamic nature of international trade law, as successive multilateral trade rounds have broadened our definition of trade distortions to encompass measures previously considered within the domestic orbit of states, or obscured by previously high tariff levels. In effect, what has changed in this view is not the degree of protection, but the definition of what constitutes protection. The increased diplomatic conflict associated with "unfair" trade practices is viewed as the result of a particularly American insistence on building a multilateral negotiating agenda that reflects its own ideological predilections, and exceeds the tolerance of our trading partners. Some NTBs are viewed as the entirely expected and legitimate substitutes for previously high tariffs as instruments in the protection of important industries, but with the considerable advantages over tariffs of lesser transparency and a porous international legal regime. Others are viewed as substitutes for quantitative restrictions, which were dramatically reduced during the previous two decades. Others, still, were undoubtedly present all along but were previously obscured by high tariff levels. Whichever the case, it is argued that the diversity of governing and developmental philosophies that characterize the major trading nations do not permit the harmonization/prohibition of key domestic policy measures through the mechanism of multilateral (or bilateral) negotiations. Implicitly, we have in this view reached the logical end of trade negotiations (at least among the major industrialized countries), and international economic diplomacy would focus on the principal causes of economic instability--namely, divergent domestic fiscal and monetary policies among the major actors, and an international monetary regime that permits long-term misalignment of key currencies and encourages sudden and drastic fluctuations in currency relationships. Indeed, if we accept this notion of inevitably divergent national development philosophies and an intractable level of protection, pressing forward with trade negotiations can have the perverse effect of replacing measures that are transparent and quantifiable in their impact with those that are opaque and contingent.[2]

A third explanation for nontariff barriers accepts the first perspective's conviction that NTBs have increased in incidence and effect. The emphasis here, however, is on NTBs as the deliberate manifestations of divergent development and strategic priorities of governments, rather than as the outcome of protectionist forces stimulated by shifting comparative advantage or by cyclical downturns in economic growth. This explanation is closely tied to the increasing concerns in this country about industrial policy and foreign government sectoral targeting. NTBs in this context are viewed as one of several policy instruments used by governments (particularly Japan, France, and Brazil) to increase output, lower the cost of production, and stimulate technological innovation in industries viewed as central to future economic growth and strategic security. NTBs enjoy within the range of industrial policy instruments a key role in protecting fledgling industries from import competition, assuring a domestic market that, when combined with other policies designed to stimulate demand, permits the development of economies of scale and lays the groundwork for the achievement of dominant market shares in both domestic and foreign markets. Again, the fact that tariffs are "bound" at low levels while the international legal regime is forgiving with respect to NTBs, as well as the opaqueness of these instruments of protection, make them particularly useful in this regard. Indeed, it can be argued that the growing appeal of bilateral reciprocity as a governing principle in U.S. trade policy is a direct result of the effectiveness of NTBs in protecting foreign high-tech markets, particularly in Japan; the recent upsurge in unfair trade (Section 301) cases brought by private parties and by the government, constitutes another reaction. It might also be argued that this effectiveness of NTBs has been the principal reason for the recent broadening of the U.S.-Japan trade agenda to include such issues as access to technology, U.S. company participation in Japanese R&D programs, protection of intellectual property and transparency of the Japanese import system.

This third explanation for NTBs is one we found most persuasive, particularly in high-technology sectors. The link between industrial targeting priorities and trade barriers is one that The Futures Group has been pursuing over the last couple of years and for which hard evidence is accumulating. In an earlier study for the National Science Foundation,[3] we cited the well-known case of copyrights on software to illustrate the development goals that nations hope to achieve by imposing NTBs on high-technology products. In this instance, the Japanese Ministry of International Trade and Industry (MITI) proposed that there be a fifteen-year statute of limitations on software copyrights in Japan, and that any company selling software in Japan be required to divulge the code containing the basic program necessary to design the package. If U.S. companies conformed to the latter proposal, Japanese competitors would obtain access to critical information that would enhance their ability to write more sophisticated software. Such information also would facilitate the jump to fifth-generation software that the Japanese believe is the centerpiece for the development of supercomputers and high-speed applications programs for business.

As a direct result of protests by U.S. industry, and pressures exerted in U.S.-Japan bilateral discussions, compulsory licensing of software has been set aside by the Japanese, and more liberal legislation proposed. The initial effort, however, illustrates the development goals that nations hope to achieve by imposing NTBs on high-technology products. Countries expect to gain greater access to the know-how that firms and high-technology industries require to improve their products and processes, know-how that may only be available from abroad. By forcing more sophisticated firms to agree to licensing and coproduction, and to offset agreements by threatening market closure, governments hope to gain detailed information about new technologies for domestic companies. Rather than barring access to their markets, nations encourage foreign high-technology firms to enter them, but only under conditions that facilitate the strengthening of domestic firms. NTBs are frequently the instrument utilized to influence the conditions of market access in ways that reinforce domestic industrial development priorities.

Such NTBs are intended to push R&D spending by national firms, to increase exports of high-technology goods, and to promote foreign direct investment by national high-technology companies. Once national firms become more familiar with new technologies, they lessen the risk involved in creating new products and processes. They can then become greater risk-takers and develop advantages in the marketplace that differentiate them from their rivals. If successful in exploiting such advantages, national firms can expand the edge they have over their rivals and create a technology gap, enabling them to overtake foreign rivals that once dominated the markets in which they compete.

The Japanese government's strategy of blocking Texas Instruments' entry into Japan's markets provides a good example of this use of targeted NTBs. Structural factors limited TI's development of a distribution network, and NTBs were used to restrict its share of the emerging local market until the firm agreed to license some of its technology. As a result, Japan gained access to know-how and design technology that it used to spur the growth of a competitive national semiconductor industry. This licensing restriction appears to have resulted in greater Japanese R&D spending, a refocusing of R&D funds on products with a greater profit potential, an improvement in export sales, and an enhancement of foreign investment potential of national firms.

In another case, the French are considering the enactment of further barriers to transborder data flows. By limiting the ability of foreign multinational firms that operate in France to use international communications, the French hope to gain access to recent research and new technologies. This could occur, since restrictions on transborder data flows would force more investment in these types of operations locally. This use of NTBs also is very different from previous ones; rather than acting to block the inflow of products, it would restrict both inflows and outflows. Such a policy would force companies wishing to do business in France to develop and store more know-how locally. In addition, keeping this information in domestic facilities will give France the ability to regulate its use.

The use of NTBs by foreign governments to target specific high-technology industries for gains in competitiveness can lead to a reorientation of U.S. companies' R&D decisions, foreign investment plans, export promotion programs and the volume and allocation of R&D spending. U.S. companies' R&D decisions can be changed if NTBs are used to target specific foreign industries. Their inability to compete in some markets because of exclusionary standards, licensing procedures, coproduction arrangements and government procurement policies can reduce the returns to R&D from products that face targeting. If returns are lower than expected, R&D funding can be reduced or R&D funds may be shifted into other areas where profits are greater and the risk of facing NTBs is lower. The need to license know-how in order to enter foreign markets lessens the control that U.S. firms have over new technologies. If this lowers returns to R&D efforts, firms may reduce R&D spending or shift R&D to other sectors.

If NTBs limit the position of U.S. high-technology products in foreign markets, U.S. firms may become much more reluctant to invest abroad. NTBs that reduce the flow of information and the ability of U.S. firms to manage a far-flung network of operations can have the same results. Barriers to transborder data flow also can have this impact. In one case, Motorola expressed its concern that French and German restrictions could force it to establish "expensive data processing facilities in each country duplicating already existing facilities" and to locate product test facilities and engineering design centers in those nations that restrict the shipment of test, certification, and design data. NCR and Xerox also have raised similar concerns because restrictions on data flows would increase their cost of doing business in Europe.

NTBs also affect export promotion programs for high-technology firms. If standards and other NTBs act to bar market entry, American firms' exports can decline. On the other hand, if foreign investments are blocked, exports may become a substitute for producing high-technology goods abroad. This can occur if licensing requirements, coproduction or joint production agreements in host nations limit the ability of U.S. companies to produce abroad. Other considerations may also lead U.S. firms to emphasize exports over foreign direct investment. These could include the loss of control over critical know-how that would result from complying with licensing and coproduction requirements.

While NTBs require sometimes major adjustments in strategy on behalf of any firm with marketing ambitions in the protected market, they have relatively greater impacts on smaller firms. This is a consequence of several factors including small firms' lack of international investment, their inexperience in international trade matters, and their lack of resources to commit in responding to NTBs. This problem is critical because small firms in a rapid growth stage need access to markets in order to regain the funds they have invested in new products and services, and to reinvest revenues in new products and markets. Participation in foreign markets is also increasingly important as a means of access to new technologies (as an increasing proportion of innovations come from foreign countries).

The lack of adequate resources is the most limiting factor in small firms' responses to NTBs. Such firms are unlikely to have staffs familiar with host country officials and regulations and will tend to spend an inordinate amount of senior executives' time in dealing with foreign (and U.S.) bureaucracies, a function normally reserved for midlevel marketing officials in larger firms. Small firms often cannot commit adequate resources to obtaining approval for products or services in cases involving standards, product certification, or licensing requirements. In such instances, the time required to gain approvals may place insupportable burdens on the firm, and may also prevent it from seizing more proximate, domestic marketing opportunities. The engagement of local legal and political assistance may be beyond the means of the company. The technical expertise necessary to demonstrate product performance for local certification bodies may be excessive. The local political influence necessary to withstand political pressures for purchases from national firms will be inadequate.

These and other disadvantages resulting from small size may force the firm to withdraw from the market and lose sales that could have contributed to its competitiveness and its long-term growth. Even when the market need not be forgone, the small firm may be required to work through a larger U.S. company that has an existing local market position, in order to gain access for its products or services. This can result in a loss of control and familiarity with the foreign market, and also retard the development of important international marketing skills in the small firm. Partnerships with locally domiciled firms may provide greater access initially, but may involve many of the same disadvantages that characterize the joint ventures with larger U.S. firms. Indeed, as is demonstrated in the case studies, some joint ventures have shown a high degree of instability over time. One further expedient that small firms often resort to in the face of market-blocking NTBs is the licensing of technology. This strategy, however, entails its own risk of invigorating a potential competitor. Depending on the specific contents of licensing agreements, this transfer of technology to local firms may often rebound to the disadvantage of the licensor in the local market, in third markets, and frequently in the firm's own home market.

Specific examples of these and other NTB efforts are contained in our case studies. The study group interviewed thirty-three companies in all, representing five high-technology industry sectors: telecommunications, semiconductors, biotechnology, computers, and software. The firms were located in California; the Washington, D.C., area; the Philadelphia area; New York; New Jersey; and the Boston area. Almost without exception, the individuals interviewed were senior or chief executives of the companies.

The firms interviewed varied widely in size. Seven firms in the telecommunications industry were interviewed, with total assets ranging from $17 million to $105 million. In the semi-conductor industry, twelve firms were interviewed; five of these firms manufactured devices, and seven made semiconductor manufacturing equipment. As a whole, this group of manufacturers' total assets varied in size from $6 million to $720 million. In the field of biotechnology, three firms were interviewed, with total assets ranging from $40 million to $55 million. The firms interviewed in the computer industry demonstrated the largest variance in size, with a range from $88 million to $2.6 billion. Total assets of the three firms interviewed in the software industry varied from under $50 million to $193 million. In addition, four other firms were interviewed: two were trading companies, and the others were in consumer electronics.

The organization of the report reflects our views on the sources/effects of NTBs in high-technology industries. Immediately following is a discussion of the principal nontariff barriers encountered by the firms we interviewed and a summary of the corporate strategies on which they had their principal effects. The remaining chapters are sectoral in focus, with introductory materials on industry structure followed by descriptions of foreign government targeting practices and general bilateral trade issues. These materials provide the context for the cases, which are also organized by sector. The concluding chapter records our observations on the trade policy implications of these empirical findings.

NOTES

1. See Robert Cohen, Richard Ferguson, and Michael Oppenheimer, <u>Nontariff Barriers to High-Technology Trade</u> (Boulder: Westview Press, 1985).

2. See Rodney de C. Grey, "A Note on U.S. Trade Practices," <u>Trade Policy in the 1980s</u>, ed. William R. Cline (Washington: Institute for International Economics, 1983).

3. See Cohen, Ferguson, and Oppenheimer, <u>Nontariff Barriers to High-Technology Trade</u>.

3
NONTARIFF BARRIERS AND THEIR EFFECTS

Nontariff barriers have gained considerable public attention over the last few years. It is now generally recognized that nontariff barriers have a significant impact on the formulation of corporate strategy in overseas markets. In particular, these barriers affect decisions concerning the location of R&D and manufacturing, investment, product design, and marketing approach. Consequently, an understanding of what these barriers are, how they work in particular markets, and how they affect the corporate decisionmaking process is essential. The companies interviewed for this study encountered a wide range of market access problems. Among those, four main nontariff barrier types emerged: standards, performance requirements, government procurement, and patents.

The most common among nontariff barriers are technical barriers to trade. Standards have become a contentious issue because of the difficulty in distinguishing legitimate from illegitimate barriers. A plethora of standards has been fostered by the wide range of concerns they address, ranging from health and safety to product quality and protection of the home market. As the rate of technological change has quickened, standards issues have taken on increased importance in world trade. In certain industries, such as telecommunications and computers, the pace of product development has exceeded the ability of major trading partners to establish commonly accepted standards, creating a chaotic situation where national technical requirements have become significant obstacles to trade. In addition to the effect of nonconforming standards, trade in some high-technology sectors has suffered from discriminatory national standards designed to protect domestic industries.[1]

American firms have found the transition to export more difficult because of the existence of nonconforming and trade-restricting standards. In addition, many of these firms themselves do not manufacture to the international standards that do exist, thus contributing to the confusion they face in world markets. This confusion is exacerbated for established technologies, such as data processing, which are not regulated by a commonly accepted international system of standards. Moreover, the International Standards Organization (ISO) on average requires five years to develop a standard. Consequently, today's rapidly developing high technology industries cannot, by definition, have agreed-upon standards.[2]

In the course of this study, the problem of nonconforming standards was raised by a large number of the companies interviewed in the telecommunications industry. Of particular concern was the incompatibility of the European CCITT standards for telecommunications equipment with U.S. standards for the same equipment. Although U.S. standards are considered to be technically more stringent than CCITT standards, U.S. companies are forced by the local PTTs to comply with these standards. These two sets of standards are sufficiently different to result in an inability in many instances to achieve compatibility between equipment manufactured to each of these standards without considerable modification. Case No. 20 highlights this issue rather well. Since Intelsat requires CCITT standards for the interface between the user and the earth station in the TDMA systems they are introducing, this company had difficulties selling its equipment for use in this system. In a few cases, the company was able to buy European-made CCITT conforming equipment and retrofit this equipment to its own system to circumvent the interface problem. However, the high cost of modifying equipment, such as TDMA, to CCITT standards prohibits the company from bidding on contracts unless there is a strong possibility of winning the contract and a high probability of long-term follow-on work on the system that would allow the company to recover the costs of modification.

This same issue came to the forefront of the semiconductor equipment industry this past year when the Japanese indicated that they were considering the establishment of a separate Japanese standards group for the semiconductor equipment industry. This was met by extreme alarm from the Semiconductor Equipment and Materials Institute, Inc. (SEMI), which has unofficially established worldwide semiconductor equipment standards since the industry's beginning. Over the past few years, SEMI has made repeated efforts to increase Japanese participation in its standards group in recognition of the importance of Japanese semiconductor equipment manufacturers in world markets. However, SEMI's efforts were stalled. It then became known that the

Japanese were positioning to set up their own industrial standards group for semiconductor manufacturing equipment. U.S. manufacturers learned that if the Japanese were to establish their own separate standards for this equipment, U.S. manufactured products would soon encounter standards barriers in the Japanese semiconductor manufacturing equipment market--a market previously unaffected by the discriminatory use of standards. In response, SEMI registered strong complaints to both MITI and members of the Japanese industry. After much negotiation between these groups, the Japanese aborted these plans for the present time.

Standards are also frequently used to discriminate against nonlocal products. This is particularly evident in the semiconductor device and telecommunications industries. This discrimination generally takes one of two forms: either there is a transparency issue in which standards are vaguely written and appear to be interpreted and applied on a case-by-case basis, or the certification process unnecessarily delays introduction of nonlocal products to the market. In Case No. 6, a custom semiconductor device manufacturer has been frustrated by the application of product standards at NTT. The company found that the NTT standards were so broadly defined that it had to send design personnel directly to the NTT central office in Tokyo to ask the particular inspector involved for his interpretation of a specification before submitting a product for approval. Unfortunately, the company has found that even these efforts, which have cost additional time and money, do not ensure that the specification will be interpreted in the same manner when the product arrives at NTT for inspection.

Interestingly, standards are used not only by regulatory bodies to discriminate against nonlocal products, but also by local manufacturers to impede sales by foreign firms. This was encountered by one U.S. company in pursuit of a large sale of semiconductor devices to a Japanese firm. Though the Japanese firm initially deemed that the company's product met its requirements and placed an order, it then began to stall the sale. By regular mail, the Japanese firm would send lists of problems with the product and request an investigation of each problem. This continued for several months until the U.S. firm's engineers finally proved that 80 percent of the complaints had been errors or mistakes induced by the Japanese firm. Shortly after that, the Japanese firm cancelled the entire order and manufactured the device itself.

The imposition of performance requirements has also become a significant barrier because of their use by governments to promote trade and local economic development. Reductions in tariffs, recently negotiated nontariff barrier codes governing subsidies, dumping, and government procurement have limited the choices available to governments to protect markets. As a result, trade-related performance requirements have become a major way to subject foreign firms with equity and management participation in domestic companies, to a variety of official policies. Performance requirements increase the contribution of nonlocal firms to the host nation's trade and industrialization goals. Since these measures were not considered in MTN discussions and are not addressed in MTN codes, they have become an important tool to effect trade patterns and investment flows. This has occurred although neither the articles nor the intent of the General Agreement on Tariffs and Trade (GATT) supports the use of government regulations as a means to discriminate against imports and in favor of domestic production.

There are two types of trade-related performance requirements: export performance requirements and local content requirements. Export performance requirements impose commitments on investing firms in a host nation "to export a fixed percentage of production or to export a specified minimum quantity of the goods produced."[3] These requirements artificially increase exports above prevailing levels and operate like an export subsidy. Local content requirements demand that a specific percentage of the value of goods produced by a foreign investor be obtained locally or manufactured locally. Such steps promote local sourcing, increase the level of local value-added, and, sometimes, provided an avenue for the transfer of valuable technology to local firms. They are similar in their impact to tariffs and import quotas. In a number of instances, these measures are not employed alone but are combined with additional requirements that can also distort the flow of trade between countries.

Performance requirements can be either explicit or implicit--in either case, the result is the same. These requirements can force a company to locate manufacturing and R&D facilities overseas, to undertake joint R&D projects with local firms, or to increase the local content of its product. In Case No. 4, a U.S. computer manufacturer found the French government relatively explicit about these requirements. The French government is very selective about which type of investment it will consider as part of a firm's necessary contribution to the local economy. For example, the company contracts out significant amounts of translation work to local firms, but the French government does not include this in its assessment of the company's contribution to the local economy. Initially, the firm began by distributing its manufacturing to local firms since this is easier than distributing developmental work and does not generally involve a significant transfer of technology. Authorities in France began to recognize, however, that although this increases local employment it does not contribute to the country's technological base. As a result, requirements have become more specific; U.S. firms are asked to perform joint hardware and software R&D with local firms and, in some cases, the type of work to be done is specified. These requirements are more prevalent in heavily targeted industries, where the country has committed itself to the development of its technological base.

Since governments are among the world's largest purchasers of goods, preferential government procurement has emerged as a major barrier to market entry: most of this vast market has traditionally been closed to foreign companies by formal and informal policies that discriminate in favor of national producers. As a result, significant comparative advantage has been lost on an international scale because of the use of less-efficient domestic suppliers. While the development or maintenance of a firm or industry may be vital in the eyes of a government, local sourcing can be a comparatively inefficient means of achieving it. Direct subsidization of the high capital costs associated with infrastructure development, for example, is far more cost-effective. Despite these inefficiencies, procurement policies are often pursued in product areas viewed as vital for both national security and the growth of other industries (electronics, computers, and software). Thus, although government procurement may raise relative costs, political goals lead to their use in the development of today's fast-growing industries and the struggle to achieve technological independence.[4]

In recognition of this barrier to trade and development, the issue of government procurement was one of the six nontariff issues addressed in the Tokyo Round. The central tenet of the resulting agreement is the extension of the national treatment obligation of the GATT to government purchases. In awarding contracts, signatory governments are obligated to give equal treatment to foreign and domestic bids, and must apply the country-of-origin rules normally used for customs clearance. The specific governmental agencies of all signatories to which the agreement applies also are identified (all state and local governments are excused from compliance). An important feature of the agreement is its insistence on transparency in the procurement process, with specific guidelines to which signatories must adhere. Finally, the Committee on Government Procurement was established to facilitate implementation and assist in bilateral disputes.

Nonetheless, preferential government procurement continues to be a major tool of government industrial targeting policy. France is a good case in point. The French government has made it clear that it will continue to support its domestic computer industry with "buy French" policies. Starting in the late 1960s, the government gave CII, a nationalized computer manufacturer, preferential treatment in procurement that extended beyond government agencies to numerous nationalized companies in various industries. This preferential treatment meant these agencies and companies were to buy from CII whenever possible; as far as can be determined, these agencies and companies paid market price for these goods. In 1973, for example, these purchases amounted to 80 percent of CII's annual sales. Today, preferential public procurement for national sources is epitomized by the French manufacturers' 63 percent share of the government's installed base (versus a private market share of only 45 percent). Preferential treatment continues; recently, after Solmer Steel Works (a French firm) placed an order with Univac, government officials sought and obtained cancellation of the order in favor of CII-HB equipment.

An even more recent case is the $156 million purchase of teaching computers (120,000 microcomputers of various brands) for the French school system. This order is the largest single purchase of computers in Europe, and part of an ambitious campaign to promote French computer literacy. Apple made an aggressive bid to win part of the sales order, and Apple President Steve Jobs went to Paris to discuss the possibility of producing the MacIntosh computer in France through a licensing or joint venture agreement. Apple's overtures, however, were countered by open hostility from French computer manufacturers, such as Thomson SA and Groupe-Bull. As a result, French officials have stated that the chief beneficiaries of the government computer literacy program will be French computer manufacturers.[5]

Case No. 3 provides a good example of these practices in the French computer industry. This U.S. company encountered exceptional difficulties competing against Machine Bull in France. About five years ago, the company was awarded a contract worth roughly $8 million to install a large-scale, multipurpose computer system in France. The customer was a state-owned manufacturing company. The contract was later retracted and awarded to Bull as a result of pressure that had been exerted from high levels within the French government. The company threatened to go to court and was awarded damages as a result. As a matter of policy, the company now refuses to quote to new customers within the French central government. Not surprisingly, the company's market share in France has remained stable over time: sales in

France have been fairly constant at approximately 10 percent of total European sales. This is inconsistent with the relative size of the French computer market. In addition, the company has been unable to partake in or benefit from the growth experienced by the French computer industry over the last few years.

Though much of the attention in this field has fallen on France, whose drive to create National Champions has led to a heavy emphasis on local sourcing, preferential government procurement also is prevalent in Japan. This may be even more significant for U.S. companies given the comparatively large size of the Japanese government market. Until quite recently, American suppliers were not allowed to bid on GOJ purchases, and the issue was considered by some to be the most damaging NTB in America's trade with Japan. These policies have been particularly effective in the Japanese computer industry. Figures from 1975 show that Japanese computer manufacturers dominated most government and government-related institutions, supplying 92.6 percent of government-related offices, 87.8 percent of local public organizations, and 89.7 percent of universities--versus a 24.5 and 23.4 percent share of the financial institutions and utility services market.

Public corporations play an important role within the GOJ market, with some 115 state agencies active in the procurement process. Many of these corporations have long followed strict buy-Japanese policies. Perhaps more significantly, these state firms have developed joint R&D programs with private industry designed to develop those products later procured by the government. The result has been a closed circle in which R&D and procurement processes are insulated from non-Japanese corporations. Foreign access to both markets and technology is thereby curbed.[6] In Case No. 10, for example, though the U.S. firm was able to obtain a landmark loan from the Japanese Development Bank to set up operations in Japan, it still has been unable to participate in MITI-funded research despite considerable efforts.

Although the Procurement Code's coverage of Japanese governmental entities is substantially broader than that for most other code signatories (and substantial progress has been made in opening up the Japanese procurement process), serious issues remain. Indeed, because of the potentially enormous sales involved (contrasted with an existing enormous American bilateral trade deficit with Japan) and the sensitive issue of competitiveness (coupled with perceived unfair Japanese practices), the issue of Japanese procurement in the advanced sectors is one of the few high-tech trade issues to emerge into the realm of high politics.

Patent protection has become increasingly important to high-technology industries that place a premium on proprietary design concepts. As a result, the slowness of the Japanese patent system has caused particular concern to U.S. companies trying to market their products in Japan. A recent study showed that the typical Japanese patent is issued about six years after the application is filed; the average time required for issue of a U.S. patent is about two years after filing.[7] In addition to the significant delays in the Japanese patent system, many companies have encountered difficulty obtaining patent protection in Japan in technological areas that are of particular strategic importance to the Japanese. This has been particularly noticeable in the fields of computer and semiconductor technology.

In Case No. 7, the firm has been unable to obtain protection in Japan for its IC devices, even though the company has patented the same technology in countries around the world. The company has several patent applications pending in Japan; it has waited as long as four years for the first action. The Japanese patent examiners have denied the company one particular patent on the claim that there is prior art in Japan. However, the company was able to obtain patents for the same technology in the United States and Germany, where it proved there was no prior art. The company appealed the findings of prior art with the Japanese patent office and was overruled. The company still has no patent protection for its technology in Japan. Fortunately, the technology is well patented in the United States, where the Japanese are planning to introduce compatible devices in the near future.

Similar problems have been encountered by U.S. companies producing computer hardware and software, an area of considerable GOJ targeting. Case No. 5 describes a company's ten-year struggle to patent its printer technology in Japan. The chief reason for denial in this case also has been that of prior art. The company applied for patents in the United States, the United Kingdom, France, Germany, Canada, and Japan in 1975. In the ten years that followed, there was no problem in obtaining adequate patent protection in North America or Europe; however, a patent for the design concept still has not been issued in Japan. The company's patent case is currently being heard by the Japanese High Court. Case No. 14 is quite similar: this company also has been struggling to obtain a patent in Japan for ten years, even though the company had no trouble obtaining identical patents in the United States, Canada, Europe, and South America.

Though the number of cases in this study does not permit statistically significant conclusions, an analysis of the strategies adopted by small- to medium-sized firms and large firms (see Table 3.1) in response to these market barriers does reveal a

TABLE 3.1
Summary of case studies

Firm No.	Total Assets ($ Mil)	Strategy	Reason
1	55	Joint venture in Japan with nonpharmaceutical company; limited technology transfer	Need for non-U.S. source of manufacture to circumvent U.S. regulations. Need for local partner to deal with Japanese regulatory authorities and system need for capital; desire to retain rights to U.S. market
2	44	Sells through Japanese trading company; technology transfer outside of company's area of proprietary research; in Europe has own marketing representative	Perceived difficulty of marketing in Japan due to cultural differences and complex regulatory structure; need for capital, but desire to maintain an area of proprietary research
3	over 500	Wholly owned marketing and service organizations throughout Europe with a manufacturing facility in Germany; two joint ventures with local Japanese companies; greater emphasis on developing products specifically for the Japanese market in the future	Nationalistic attitudes throughout Europe; "buy national" in France; difficulty selling to PTTs in Europe; teamed with local Japanese companies that were not competitors; heavy competition in large-scale computers from local Japanese companies and appearance of regulation in that market
4	502	In Europe and Japan, began with distributors and then set up wholly owned subsidiaries for sales and marketing; also wholly owned manufacturing subsidiaries in Europe	Increasing pressure to contribute to local economy; pressures to perform local R&D, joint R&D with local firm, or source from local companies; difficulty retrieving some of the costs associated with adapting the product to Japanese market
5	88	Licensed product in Japan and lost market there	Need for capital; did not do sufficient R&D to create products for local Japanese market; unable to obtain patent in Japan
6	over 100	Joint venture in Austria for manufacturing and design; Design Center in Japan	Circumvent duty system; enable interactive custom design; difficulties selling to European PTTs (specifications used to favor local vendors); broadly defined NTT specifications; nationalistic attitudes
7	96	Joint venture in Japan since the 1950s for sales and marketing only (not very successful); hope for wholly owned manufacturing in Asia in the future; no great optimism for European market	Unable to obtain patents in Japan; problems with "slippery" standards in Japan and Europe; nationalistic attitudes in Europe and preference for Siemens in Germany

(Continued)

TABLE 3.1 (Cont.)

Firm No.	Total Assets ($ Mil)	Strategy	Reason
8	265	Wholly owned sales and manufacturing subsidiaries employing local nationals	Strong commitment to "making it on its own" and desire to maintain control
9	25	Joint venture in Japan for sales, manufacturing, R&D for local market; wholly owned subsidiaries in Europe	Problems with distributor relationships in Europe (copied product) and Japan (no contact with customer); difficulty obtaining financing for 100 percent owned facility in Japan; received pressure from French government to form a joint venture with local firm and transfer technology
10	65	Began with trading company in Japan and later established joint venture for sales and manufacturing; wholly owned sales subsidiaries in Europe with manufacturing site in France	Need to have more direct involvement in Japan because of market size and importance; difficulty hiring salesmen and engineers in Japan; unable to participate in MITI-funded research
11	6	Licensing agreement in Europe that resulted in a breach of contract; licensing agreement in Japan that had to be terminated; currently using distributor in Japan	Loss of technology through licensing agreements in Europe and Japan
12	300	Originally used trading company in Japan for distribution and service; in 1980 established joint venture in Japan for service, with intent to develop sales and manufacturing there; in Europe sold through distributors until 1973, then established a sales and later manufacturing subsidiary in Europe; in 1981, established joint venture in France for R&D manufacturing and sales	MITI-funded development of competing technology given to Japanese firm and Japanese customers urged to buy from this local company; Japanese targeting of domestic industry; Japanese customers began to require customization of product that could not be provided from U.S. manufacturing site
13	190	Joint venture in Japan that resulted in loss of technology; after this experience, used a local distributor in Japan until recently, then a joint venture for sales and manufacturing was established	Loss of technology through a joint venture in Japan; as Japanese catch up technologically, company must provide more and more modification of product; inability to hire work force unless affiliated with large company in Japan

(Continued)

TABLE 3.1 (Cont.)

Firm No.	Total Assets ($ Mil)	Strategy	Reason
14	1	Delayed entry to Japanese market	Inability to obtain adequate patent protection in Japan
15	200-300	Joint venture with local partner in two European countries and Japan; joint R&D in these countries with R&D also at headquarters	Need to look like a local company; need for understanding of local client; EC procurement policies obscured information for nonlocals
16	150	Distributors around the globe with wholly owned sales subsidiaries in key countries; wholly owned manufacturing and R&D facility in Europe; joint R&D in Japan	Need to show commitment to the British market to sell to nationalized buyers; need to work with a local Japanese company to properly transform product
17	10-15	In Europe, began with distributors and slowly replaced with joint ventures for sales and service in Japan, work with trading company	Need to work closer with customer; difficulty selling to NTT; no appreciable amount of business with European PTTs; difficulty selling to Japan because of Israeli manufacturing site
18	65	In Japan, works with value-added reseller to which it provides all sensitive supporting documentation such as specifications and source code in exchange for worldwide rights (excluding Japan) to market any enhancements to the base technology developed by the Japanese reseller	Local firm helps in certification of company's product in Japan; perception of improved service capabilities in local market; commitment to sell to NTT
19	85	Originally had distributor in Europe, but terminated the arrangement after it learned that distributor had copied technology; later wholly owned sales and service subsidiaries and manufacturing facilities in Europe	European distributor copied technology and sold its own competing product; needed in-country presence to be considered a reliable supplier; service is key; "buy national" in France; pressure to manufacture locally in France; difficulty with German PTT and qualification testing; incompatible CCITT standard in Europe; difficulty with NTT standards
20	105	Uses own U.S.-based sales force to sell overseas	Difficulties due to CCITT standards used in Europe; preference for local companies shown in government procurement; need to bid through local

(Continued)

TABLE 3.1 (Cont.)

Firm No.	Total Assets ($ Mil)	Strategy	Reason
20 (Cont.)			company in Germany; perceives Japanese market as completely closed and does not pursue
21	63	Sells through an international distributor; in Japan, selected an individual to act as agent	In France, difficulty selling to PTT, which became more severe once its local distributor was nationalized; in Germany, received pressure from the Bundespost to transfer technology to several local firms; overall, nationalism a problem in Europe; certain limits to the distributor relationship when times are bad; in Japan, perceived the market to be closed during the 1970s and did not attempt to enter, in the 1980s adopted a cautious approach
22	under 50	Uses distributors	In Germany, Bundespost has used standards to urge company to work with local company, also problem with export of sensitive data; problem of incompatibility with CCITT standards in Europe
23	under 50	Joint ventures with local partners	Uses local partners to deal with local regulatory authorities; some problems with German Bundespost which uses standards to manipulate terms of doing business; "buy national" in the United Kingdom, Germany; opposition from the KDD was overcome by local Japanese partner

pronounced dichotomy that, to a large extent, reflects the gap in the resources of these two groups. Large companies are more familiar with the intricacies of foreign markets and will not, in general, go to the U.S. government for assistance in penetrating a foreign market. These firms tend to have greater bargaining power in the foreign marketplace due to the large number of jobs and the size of the investment they can generate. Thus, large firms have a wider range of options available to counter market barriers. Smaller firms, on the other hand, generally do not have substantial export experience nor large sources of capital to invest in risky ventures. As a result, their options are greatly limited. An examination of the case studies provides useful insight into the different strategies adopted by these two groups. Interestingly, there are certain parallels between the strategies adopted by the large firms early in their history and the strategies adopted more recently by smaller firms--in twenty-five years, little has changed.

A small- to medium-sized firm that confronts trade-restricting standards in an overseas market is likely to seek the representation of a domestic firm to deal effectively with local regulatory bodies. Within this group, the smaller and less-experienced firm, for example, will opt for a relationship with a local distributor or trading company that is familiar with the local regulatory system and can ease the entry of the firm's product to the market. This strategy was adopted by the firm in Case No. 2, for example. This young biotechnology firm never seriously considered any other option because of the complex regulatory processes involved in the commercialization of biotechnology products. However, examples of this are more common in the telecommunications industry due to the frequent use of standards by the local PTTs to discriminate against nonlocal products and the complex certification processes throughout Western Europe and Japan. Thus, examples can be found in Cases No. 17, 18, 21, and 22. In each of these cases the company chose to be represented by a local trading company or distributor because it perceived that the local firm would be able to deal more effectively with standards and product certification procedures. In Case No. 18, for example, the U.S. firm's Japanese distributor successfully applies for certification exemption on a case-by-case basis--a cost- and time-saving procedure previously unknown to the U.S. firm.

When confronted by these same standards issues, a larger firm will use its considerable resources to negotiate a way around the barrier through its local sales subsidiary, or will establish a production facility within the market and manufacture to local specifications. For example, though the firm in Case No. 19 began with a distributor in Europe, it eventually set up its own wholly owned sales and service and manufacturing subsidiaries in Europe. Large firms are more successful at dealing effectively with these standards problems because they have the resources to establish staff and operations in foreign markets employing local nationals who are familiar with the practices of the host nation, have contacts within the local regulatory system, and can react quickly to any problems that arise during the certification process. Unfortunately, the large capital expenditures associated with this strategy militates against its use by smaller firms.

The reaction of small and large firms to the imposition of performance requirements did not vary that greatly, though some difference can be noted. Both sets of firms are wary of licensing technology to local firms in response to government pressure. For example, the small company in Case No. 9 waited a year and a half to establish a wholly owned subsidiary in France because of pressure from the French government to form a joint venture with a local firm. Similarly, the large company in Case No. 4 has not to date been involved in any substantial joint R&D efforts in France despite persistent requests from the French government. In instances such as these, the government's intent to boost its local technological base is obvious, and therefore both groups of firms recognize the potential for significant technology loss. However, larger firms, because of the greater resources they command, have more options available to them in satisfying a local government's request. When possible, these firms will try to source or manufacture locally, rather than transfer any significant technological know-how. Moreover, the large firm tends to be in a better bargaining position vis-a-vis the local government because of its potential to generate significant local employment and export revenues. Thus, if the large firm does choose to work with a local firm, it is able to sufficiently narrow the legal definition of that arrangement to preserve its proprietary know-how and avoid the emergence of direct competition from this relationship.

Preferential government procurement can cause a smaller firm to avoid the affected market completely. In Case No. 20, the closed nature of NTT procurement caused the firm to dismiss that potential market altogether. Case No. 21 is similar; for the first ten years of business the firm did not even try to sell into the Japanese market. This suggests a tendency among smaller firms to avoid a market that they perceive to be heavily influenced by "buy national" policies and to concentrate efforts elsewhere. For the most part, a larger firm will try to manufacture within the country or otherwise contribute significantly to the local economy. In this way, the large firm makes it more difficult for the local government to discriminate against its product. This is largely the motivation of the firm in Case No. 19 to consider establishing a manufacturing facility in Japan. Even some large firms, however, have had such negative experiences resulting from preferential government procurement (Case No. 3) that they have opted to avoid the affected market entirely.

Finally, patent issues can affect small and large firms differently. The small firm in Case No. 14 chose to delay its entry to the Japanese market until its patent dispute is settled. This firm's viability in the long run is significantly influenced by the success of this particular product in the Japanese market. Consequently, the firm does not want to risk having its technology copied by the Japanese with no legal recourse. Large, more mature firms as a rule have more diversified product lines and are not dependent on a single technology for their success. Further, these firms have more resources to allocate to R&D projects and are constantly developing new products. As a result larger firms confronted with the inability to obtain adequate patent protection in a market may enter that market despite the lack of protection and innovate as quickly as possible. This strategy is used by the firm in Case No. 7. Here the firm stressed that the key to its success was to stay ahead of its competitors in terms of technological innovation--thus, by the time the firm was introducing a product to market, it already had another, more advanced version near production. Therefore, the larger firm aims to obsolete its own invention and will not slow down the innovation process because of a lack of patent protection.

The consequences of these strategies can be significant. Certain conclusions concerning the efficacy of the strategies adopted can be drawn from an analysis of the case studies. These strategies ranged from licensing agreements with local firms, to selling through a trading company or local distributor, forming a joint venture with a local firm, and setting up a wholly owned subsidiary in the foreign market. Regardless of the size of the firm implementing a given strategy, some strategies are more prone to complications than others.

Overall, the licensing agreements encountered in this study did not prove to be effective over the long run. In a number of cases (Cases No. 9, 11, and 19), the licensing of technology to a local firm in a foreign market led to the loss of technology in both the European and Japanese markets. Companies found that the licensing arrangement was too difficult to control from a U.S. base. Consequently, the arrangement developed in an unwieldy manner and led to the licensee's development of a competitive product which it would then sell directly into its domestic market. In most instances, because of the geographical distances involved and the licensor's lack of familiarity with the overseas market and customers, the licensee was able to develop this competing capability without the knowledge of the licensor. In Case No. 19, for example, it was not until an overseas customer wrote to ask why equipment supplied directly from the U.S. firm was different from the equipment it had previously received from the licensee that the firm realized that its licensee had copied its technology. In the majority of these cases, the U.S. firm has chosen not to seek legal remedies because of the time and money involved in pursuing such a case in a foreign country. Instead, the licensing agreements were terminated and new strategies were developed. Case No. 2, however, provides an interesting example of one small- to medium-sized firm that has effectively managed licensing agreements. In this case, the firm has been successful because it is able to license technology in selected areas while maintaining certain niche product markets for itself. For small companies in a position to segment their R&D and product development in this fashion, this can be an effective means of circumventing market barriers, while increasing overall revenues.

Many companies retain a local trading company or distributor because of limited funds available for the exploitation of overseas markets and unfamiliarity with foreign regulatory bodies, business practices, and consumer preferences. However, over time, these relationships become less satisfactory because they insulate the U.S. company from its overseas customers and retard the firm's ability to react to developments in the overseas market. In addition, important international marketing skills are not developed. Consequently, as the overseas market becomes more important to the firm, the distributor relationship becomes a less-effective means of reaching that market and the firm begins to consider a direct sales presence. This scenario is particularly relevant for the telecommunications industry where the perception of a service capability is critical. Cases No. 17 and 21 provide examples of how this relationship can become restraining. In Case No. 17, the firm used local distributors in Europe until its European business grew large enough to support a direct presence in the marketplace. In Case No. 21, the firm used an international distributor and found that its products tended to do well only in countries where the distributor did not have a competing product of its own. In these cases, the distributor relationship is an effective means for entering a market, however, not a viable strategy over the long run.

On the whole, joint-venture agreements with the local firms have proved to be an effective means of penetrating a market blocked by preferential government procurement, standards problems, or significant performance requirements. However, these arrangements were shown to be more stable over time if the provisions for the transfer of technology and rights to the use of that technology by the joint-venture partners in world markets were narrowly defined in legal terms, and if the local partner was not a potential competitor of the U.S. firm. In Case No. 13, the U.S. firm incurred a serious loss of technology as a result of a joint-venture arrangement in Japan that did not have sufficient legal safeguards. In this case, the Japanese joint-venture partner did have business interests that overlapped those of the U.S. firm prior to the establishment of the joint venture. Other joint ventures have, however, fared rather well. Case No. 3, for example, describes two Japanese joint ventures in the computer industry. This U.S. company consciously chose to team up with Japanese firms that were not potential competitors and thus would view the arrangment as a profit-generating venture rather than a way to gain needed technological know-how. This company's decisions have yielded two Japanese joint-venture arrangements that are profitable and have remained stable for over twenty years.

Several of the smaller companies interviewed have successfully implemented this strategy (Cases No. 9, 10, and 17). In Case No. 9, the firm's approach was to establish joint ventures overseas with local individuals with whom the firm had already established a trusting relationship through previous business transactions. Consequently, the firm was able to create autonomous overseas business ventures that benefit from local part-ownership. These ventures are profitable for the overseas partner and, therefore, there are no competing interests involved and there is no incentive to steal technology. Further, this type of joint venture, initially only for marketing and sales, requires little capital to start up. Overall, the study team found this strategy to be the most successful for small- to medium-sized firms.

The most effective means of circumventing market restrictions has, however, been the establishment of a wholly owned subsidiary in the overseas market. The establishment of a wholly owned subsidiary has allowed the U.S. firm to develop a critical understanding of the target market and local regulatory processes through the hiring of local nationals and proximity to the market. In this way, the firm is able to successfully confront standards problems. Through the establishment of a wholly owned subsidiary, a firm is able to satisfy performance requirements with greater ease and fewer compromises to overall corporate goals. Moreover, as the firm's contribution to the local economy through employment and capital investment for facilities increases, the firm bypasses market-restricting preferential government procurement and, in fact, may even benefit from these policies. Though for the most part this strategy is adopted by mature companies that have considerable financial resources, Cases No. 9 and 10 indicate that smaller firms should consider this option if the potential market is large enough and risk-taking can be adequately managed.

40

NOTES

1. Robert E. Baldwin, Non-Tariff Distortions of International Trade (Brookings Institution, 1970).
2. From interviews with officials at USTR, April/May 1984.
3. The Labor-Industry Coalition for International Trade (LICIT), Performance Requirements (Washington, D.C.: LICIT, March 1981), p. 2.
4. Abel Farnoux, Excerpts from the Report by the Committee for Electronics (Paris: French Ministries of State and of Research and Technology, March 1982).
5. "Apple Computers vs. French Chauvinism: Politics, Not Free Trade, Wins in the End," Wall Street Journal, March 1984.
6. USTR, Japanese Trade Barriers to U.S. Trade and Recent Japanese Government Trade Initiatives (November 1982). Electronics Industry Association, "Changes in U.S. Telecommunications Industry and Impact on U.S. Telecommunications Trade," Submission to the International Trade Commission, April 17, 1984.
7. U.S.-Japan Trade Study Group, "Progress Report: 1984," September 1984, pp. 57-58.

4
BIOTECHNOLOGY

The biotechnology industry is still in a nascent stage. The industry is involved in everything from brewing beer and treating sewage to genetically engineering bacteria and embryos. The field, however, is undergoing a transition from what was mostly a research endeavor to what is now a serious commercial and international undertaking. This transition is expected to narrow the seemingly disparate activities encompassing biotechnology toward meaningful industrial and health applications in sectors such as agriculture, pharmaceuticals, and chemicals.

It is estimated that 850 firms are performing biotechnology research, of which approximately 220 biotechnology firms, or 30 percent of those companies, are domiciled in the United States. Of these, roughly half are newly formed genetic engineering firms. Over one hundred of these new biotechnology firms (NBF) were formed between 1976 and 1984, including forty-three in 1981 alone.[1] Many of these companies were formed around a core group of university scientists and began by granting R&D contracts to academic researchers and university groups for commercial development of discoveries made in university laboratories. This group includes such firms as Genentech, Centocor, Genex, and Amgen. The other half of these firms are larger, established companies from the pharmaceuticals, food, energy, and chemical industries that undertake biotechnology research to improve on existing production processes or for new product development. An example of one such company is Monsanto Corporation. Monsanto currently is investing in a new $150-$200 million Life Sciences Research Center that will eventually house 1,200 scientists and technicians who will study chemical and biological characteristics that control the growth, development, and health of plants and animals. Through these efforts, Monsanto hopes to produce new crop chemicals, nutritional compounds, and pharmaceuticals to keep its business growing through the 1990s.[2]

41

Industry analysts have identified five classes of new biotechnology firms. These are: therapeutic companies, which develop products designed to treat disease; diagnostic companies, which make or develop products to diagnose disease; therapeutic/ diagnostic companies, which make or develop products in both the aforementioned categories; scale-up industrial processing companies, which develop commercial scale processes or products for industrial use; and agriculture/animal health care companies, which develop products to improve crop yields or benefit animals.[3] According to a recent OTA study, 62 percent of all biotechnology firms and 72 percent of the new biotechnology firms are pursuing projects with applications in the pharmaceutical industry which would include both therapeutics and diagnostics applications. Twenty-eight percent of all biotechnology firms are involved in animal agricultural applications, and 24 percent in plant agriculture.[4]

The growth in the number of new biotechnology firms in the late 1970s and early 1980s was spawned primarily by the willingness of the U.S. venture capital markets to invest in young biotechnology start-ups, and the willingness of talented scientists to leave the protected bastions of academia for the more lucrative private-business sector. Most of these start-ups financed their own proprietary research by performing contract research for large U.S. companies (many of which belong to this other established half of the biotechnology industry) and foreign companies. This contract research tended to focus on new product development in return for which the NBF received needed cash and royalty income from eventual sales. Of the fifty-seven arrangements between small and large U.S. biotechnology firms cited in a recent study, roughly 40 percent involved an equity share in the small firm by a large biotechnology firm. Almost one-quarter of these fifty-seven arrangements involved the formation of a joint venture for research and development.[5] Clearly, the financial ties between these two sectors of the U.S. industry are substantial. There are two main reasons for the proliferation of these arrangements: the inability of the small biotechnology firms to commercialize and market products based on their research; and the failure of the large, established firms to recognize early on the significance of genetic engineering and its potential applications in industry.

As these companies began to feel the need to concentrate more heavily on their own proprietary research, resources started to shift away from contract research. As a result, most of these firms are now generating needed funds through public offerings, private placements, and limited R&D partnerships. Limited R&D partnerships, for example, raised $148 million and $160 million in financing for NBFs in 1982 and 1983 respectively.[6]

Although small upstart firms have clearly led the way in the field of biotechnology in innovation and technological break-throughs, they have been aided by the larger, more established firms that have provided needed capital and sources of revenue. Small companies constitute roughly half of the U.S. firms currently engaged in biotechnology research. Industry analysts, however, do expect a shakeout to occur over the next eighteen months. Of the nearly fifty public U.S. biotechnology companies, only a half-dozen have shown any profit to date.[7]

The tie between academic institutions and private business has been very important to the development of the U.S. biotechnology industry. University research has been particularly impor-tant in the fields of enzymology and molecular biology. The best example of this important link is found in the Cohen-Boyer patent for gene-splicing, which belongs to Stanford University. There are several types of relationships between universities and indus-tries in the United States that include consulting arrangements, industrial associates programs, research contracts, research partnerships, and private corporations. These relationships are considered to be very successful in transferring technology to the private sector in the United States; similar relationships in Japan, the United Kingdom, West Germany, and France are not as well developed, although emphasis is now being put on these relation-ships.

Currently, the United States is the leader in the field of biotechnology in terms of innovation through genetic engineering and basic research. However, since the industry is just beginning to enter the period of commercialization it is still too early to gauge whether the United States will maintain its lead. An OTA report recently identified the five countries most likely to challenge the U.S. preeminence in the field of biotechnology as Japan, West Germany, France, the United Kingdom, and Switzerland.[8] An inspection of the list of U.S. patents granted in six areas of biotechnology products and processes gives indication of the competitive state of U.S. research and development in that field. Of the 919 U.S. patents granted in these six areas between 1980 and 1982, 55 percent were issued to U.S. companies, 20 percent to Japanese companies, 6 percent to West German companies, 5 percent to U.K. companies, and 3 percent to French companies.[9] These figures suggest that the United States has a substantial lead in biotechnology.

In terms of the quality of basic research, the United Kingdom is regarded as second to the United States. In the past, most of the R&D in biotechnology was performed by large, established companies because the availability of venture capital is very limited in the United Kingdom. Small research firms, however, are beginning to appear. These include Celltech, Agricultural Genetics, Plant Sciences, Imperial Biotechnology, and IQ (Bio), which were formed specifically to exploit the results of basic research. The most successful of these is Celltech which admittedly had modeled itself after Genentech, the leader among the new U.S. biotechnology firms. Fifty-six percent of Celltech's initial funding came from private financial institutions and 44 percent from the British government. Celltech, established in 1980, has negotiated joint ventures with Boots and Air Products for substantial down payments. It is the world's leading producer of anti-interferon, which is used primarily to purify interferon. The firm's commercialization efforts have been relatively impressive thus far.

The major role in the commercialization of biotechnology will be played by the large, established U.K. firms such as Burroughs-Wellcome, Unilever, ICI, G. D. Searle, Glaxo, and others. Most of these firms have already begun extensive R&D programs in-house. These companies are well equipped to handle the high costs associated with the commercialization of biotechnology products that come from large-scale production, certification of products, and marketing.

The biotechnology industry has been the subject of substantial government targeting efforts since a paper written by the Royal Society, the Advisory Board for the Research Councils, and the Advisory Council for Applied Research and Development (ACARD), referred to as the Spinks Report, was published in 1980. This report showed that the impact made by private firms to the biotechnology industry was insufficient and that government assistance on a fairly large scale was needed if the U.K. biotechnology industry was to remain internationally competitive. The instrument of the government's subsequent policy includes support for basic and developmental research; promotion of joint ventures between academia, private firms, and the government; and some partial funding of commercial ventures.

Although there are a number of government departments involved in supporting the biotechnology industry, overall coordination is provided by the Department of Trade and Industry (DTI) through the Biotechnology Unit which was established in November 1981. This group is composed of experienced individuals from industry who advise on the allocation of resources. The other major force is the Biotechnology Directorate established by the Science and Engineering Research Council (SERC), which coordinates research grants and helps to identify potential growth areas.

In November 1982, the DTI launched a three-year program to help the development of biotechnology. Since the United Kingdom is relatively strong in basic research and weak in commercial applications of biotechnology, the government's efforts are focusing on projects that will help firms to bring products to market. This program will spend 16 million pounds to support basic research, problem-solving demonstration projects, feasibility studies, strategic studies, and industry infrastructure development. The program will fund up to 50 percent of strategic studies projects (maximum of 25,000 pounds), 100 percent of problem-solving projects (cost of five to fifteen days of consulting), and 75 percent of feasibility studies (maximum 10,000 pounds).[10] Through these projects, the DTI encourages and supports biotechnology within companies. Supported projects cover a wide range including the large-scale production of monoclonal antibodies at Celltech.

SERC's Biotechnology Directorate estimates it is spending about $4 million per year. Firms must contribute at least 50 percent of the funding to SERC projects. Among these projects was the formation of a genetic engineering research center at Leicester University and several companies to exploit the results of university research which include Plant Sciences, Ltd., Imperial Biotechnology, Ltd., and Bioscot. SERC also has encouraged the development of advanced courses in biotechnology and increased its support for PhD students.

The British Technology Group (BTG) also has aided the biotechnology sector. This group provides financing for joint ventures between industry and government. Typically the BTG provides funds for research, and private companies pay operating expenses and market any commercial applications. The most successful of these BTG-funded companies is Celltech which received 44 percent of its original funding from the group. As of July 1984, the BTG had made forty investments in this industry worth roughly $20 million.

The French biotechnology industry currently is not in a good competitive position against its American, European, and Japanese competitors. This deficiency is believed to reflect the grave shortage of qualified French research personnel in this field. The French biotechnology industry, composed of roughly sixty firms, is dominated by large, established companies, a number of which are either wholly or partially state owned such as Roussel Uclaf, Elf-Aquitaine, and Rhone-Poulenc. Though there are a few small, newly established companies, most of these are subsidiaries of the large firms or companies associated with the research institutes. The French industry does have some strength in a few small market niches such as amino acids, but overall the French lag behind the United States, the United Kingdom, West Germany, and Japan in the commercial development of biotechnology.

The French government began targeting the biotechnology industry in early 1981. A mobilization plan entitled "Programme Mobilisateur--L'Essor des Biotechnologies" was developed specifically for the biotechnology industry and was presented by the Biotechnology Mission of the Ministry of Industry and Research in July 1982. Through this program, the French hope to capture 10 percent of the world market by 1990. This plan outlined a ten-year program for development of the domestic biotechnology industry using a multidisciplinary approach that included educational development, industrial development, technology transfer, and increased funding of research. The plan forecast government spending at $81 million and private industry spending at $135 million per year between 1983 and 1985.[11] These funds are being funneled into projects that involve the strengthening of basic research, but emphasize research that will lead to the commercialization of genetically engineered products such as cell cultivation, bioreactor development, and separation and purification techniques. Attention also is being focused on those areas where the French already have a demonstrated strength such as immunology or a substantive need such as fuel additives and fertilizer substitutes.

Another part of the French government's general strategy toward the biotechnology industry was the adoption of a general research law (Loi de Programmation et d'Orientation) in July 1982. The objective of this law was to guarantee a real annual increase in the overall civilian R&D budget, to set up seven technological plans (of which the above biotechnology plan was one), and to encourage scientists and research institutes to work in collaboration with private-sector colleagues and to teach in universities.[12] It was hoped that these measures would promote the necessary multidisciplinary approach needed to bring the French biotechnology industry up to the standards being set in the United States and Japan.

Additionally, four technology centers were founded to help facilitate the transfer of technology from the research laboratory to industry in the areas of microbial engineering, fermentation, biochemical engineering, and process engineering. The French government is encouraging the formation of joint research arrangements with other countries to gain knowledge in production technology.

There are approximately sixty companies performing research in the field of biotechnology in West Germany. The field is dominated by large pharmaceutical and chemical companies. This group includes Hoechst, Bayer, BASF, Merck, Boehringer, and Schering. All of these companies have cooperative arrangements with domestic and foreign research institutes to supplement their own in-house resources. West Germany's future strength lies largely in the substantial resources for commercialization of products of this group. Hoechst, Bayer, and BASF are three of the four largest chemical companies in the world. During the last four years, several small research-oriented companies have been established including Genbiotech GmbH, Organogen GmbH, and Progen Biotechnik.[13] Thus far, small- to medium-sized firms as a whole have not played a major role in biotechnology innovation. Given the general lack of venture capital in West Germany and the preference shown by the Risk Financing Society (Deutsche Wagnisfinanzierungs-Gesellschaft, a consortium of twenty-eight banks formed in 1975 to fund high-risk start-ups) for projects that come to fruition within a two-year period, it is unlikely that these firms will play any significant role in the near-term future.[14]

The West German government began to formally target the biotechnology industry in 1979. The Federal Ministry for Research and Technology (BMFT, Bundesministerium fur Forschung und Technologie) set forth a plan for the biotechnology industry that year that gave very specific objectives for research and research applications. More recent discussions have centered on specific product areas and processes, specifically pharmaceuticals and plant agriculture. The BMFT supports biotechnology through three types of programs: funds set aside specifically for biotechnology, grants going to biotechnology as part of existing industrial development programs, and funds given through other organizations as part of support programs for public research. It is estimated that combined BMFT funding in 1982 was between $50 million and $70 million.[15]

Japan has a long record of success in the application of conventional techniques of microbiology in industry. Japan's successes in this field include the mass production of a variety of antibiotics, vaccines, pharmaceuticals, amino acids, enzymes, and food additives. In certain areas--for example, the use of amino acids in fermentation--Japanese industry is judged to be preeminent. In other areas, such as genetic engineering, although it is rapidly advancing, Japan presently lags the United States by a significant margin and often must rely on licensing agreements with overseas firms for new products and technology. In recent years, however, biotechnology has emerged as an important focus of industry and government-sponsored research in Japan. In 1981, for example, MITI selected biotechnology as a major theme in its "Basic Industrial Technology Research Program." Thus, the existence of a variety of ambitious research programs, together with Japan's record of success in the commercial exploitation of new technologies, has sparked serious concern among Japan's competitors in regard to the medium- and long-term potential of its biotechnology industry.

Of the 12,000 scientists in Japan working in the field of microbiology, approximately 70 percent are working for private firms. According to a 1983 survey, over 150 Japanese firms are presently engaged in biotechnology R&D. These companies differ from typical U.S. biotechnology firms in two major respects. First, the Japanese companies are typically large, well-established firms such as Suntory & Mitsubishi Chemical that have diversified into biotechnology in search of new growth opportunities. Second, firms from a very broad range of industries, including the pharmaceutical, food, electronic, beverage, and chemical industries, have become involved in the field. However, in terms of aggregate R&D funding, surveys indicate the U.S. biotechnology firms outspent their Japanese counterparts by a ratio as high as 10:1 in the early 1980s ($500-$700 million versus $50-$70 million).

Government support for biotechnology research and development in Japan is concentrated in three agencies: the Ministry of International Trade & Industry (MITI); the Science & Technology Agency (STA); and the Ministry of Agriculture, Forestry & Fisheries. The Ministry of Health & Welfare (MHW), Ministry of Education, and the Environmental Protection Agency also sponsor biotechnology-related R&D, reflecting the high level of government interest in biotechnology as well as its broad application potential.

Low-level public-sector support for biotechnology R&D in Japan began in the early 1970s, but did not reach significant levels until the early 1980s. Although the GOJ has designated biotechnology as a key basic technology, the government sponsored share of biotechnology R&D in Japan remains relatively low--an estimated 20-25 percent of the total. Furthermore, the bulk of these funds is allocated to government research laboratories and regulatory programs and does not provide direct assistance to private-firm R&D efforts. However, the GOJ has sponsored some ambitious long-term R&D efforts, the most important of which are described briefly below.

- The Technical Research Association for the Development of Biotechnology, consisting of fourteen major pharmaceutical, chemical, and food companies, was established in 1981 under the auspices of MITI's "Basic Industrial Research Program." Over the next decade, this organization will be responsible for allocating 26 billion yen on biotechnology research in the areas of biochemical reactors, mass cell culture, and recombinant DNA.

- Three recent projects sponsored under the Japan Research & Development Corporation's Exploratory Research for Advanced Technology program are focusing on long-term R&D in the areas of bioholonics, bioinformation transfer, and microorganisms that can grow under extreme environmental conditions. Each project will receive total government funding of some $7.2 million.
- In 1979, the STA launched a major program focusing on research in the areas of recombinant DNA, genetics technology, and bioreactors. The program ran for four years, with a budget of approximately 1.5 billion yen per year.
- The MAFF established a National Institute of Genetics to perform research in genetic engineering.
- The STA's Science Council supports biological research in about 100 universities.
- In 1975, MITI classified microbiology as a "priority technology." At that time it spent approximately 6 billion yen on roughly 200 microbiology-related projects.
- Under Aqua Renaissance 90, a new MITI-sponsored "national large-scale technology project," $52 million will be spent over the 1985-1991 period to develop a bioreactor capable of processing and purifying wastewater.

Other types of government support for the biotechnology industry in Japan do not appear to be significant. No special tax treatment is provided to the industry, although it does benefit from the general tax breaks given to R&D activity and to firms that are members of MITI-approved research associations. Also small quantities of low interest loan funds are available to firms through the Japan Development Bank and MITI's Venture Enterprise Promotion program.

Since commercialization of biotechnology products has only just begun and the effects of government targeting, now in effect since 1981, have not yet been felt, a discussion of potential trade issues in this nascent industry is purely speculative. Nevertheless, an examination of industry patterns in related industries, current targeting measures, and industrial developments gives some indication of future trends and important issues.

Patent regulatory issues are certainly going to play an important role in the future because of two factors: the general lack of experience in applying patent laws to biological systems and products, and the differences in patent systems among nations. Patent issues are already coming to the forefront of the

U.S. biotechnology industry. For example, there is a wide debate over the patentability of selected products. Some claim that most potential bioproducts, such as interferon, human growth hormone, and human insulin, are "products of nature" and, therefore, are not new. In addition, there is considerable debate over what makes one life form distinct from another--some genetic inventions differ from a competitor's only in the composition of a string of two dozen amino acids out of hundreds. These disputes are particularly contentious because they involve some of the broadest and potentially most lucrative patents. One such case is the Biogen-Genentech dispute that began in February 1984 over the exclusive rights to market alpha interferon in Europe. This case is important because it will determine which company will be able to say that it was the first to market interferon and will also sharply affect the fortunes of two major pharmaceutical firms, Schering-Plough and Hoffmann-LaRoche, who are acting as marketing agents for Biogen and Genentech, respectively, and have invested millions of dollars into the development of the products.[16]

Foreign patent laws significantly differ from U.S. patent laws. In the United States, a patent application is held in secrecy until the patent is issued; in Western Europe, public inspection of a patent is allowed for eighteen months after filing regardless of whether the patent is granted. In the United States, a written description of the invention may be made publicly available up to one year before the filing of a patent application; in Japan, the grace period is six months; in Europe, there is no grace period. In the United States, if there are several separate applicants (each of whom must be the inventor) for the same invention, priority goes to the first to invent. In Europe and Japan, priority goes to the applicant (who does not have to be the inventor) who was the first to file for a patent application. In addition, the European Patent Convention (EPC) does not consider new plants or animal types, nor surgical and therapeutic methods that involve the treatment of the human or animal body for patent protection; Japan will not consider inventions that involve unclear processes or methods for patent protection.[17] Finally, considerable differences exist between the United States, Japan, and Europe over the issue of what constitutes patent infringement. All the differences outlined above indicate that there is fertile ground for international patent disputes in the field of biotechnology. Given the importance of these patents to the companies involved, it is expected that patent issues will become an instrument of government targeting policies in the future.

As foreign competitors begin to catch up with the U.S. biotechnology industry it is expected that the number of contract R&D arrangements between U.S. and foreign firms will decline. As the competition for these contracts increases, requests from customers for more significant technology transfer are very likely to increase as the customer gains greater leverage and bargaining power. In Japan, each of these R&D contracts must be approved by the Japanese government. Thus, U.S. companies may encounter difficulty receiving this approval in areas where the Japanese industry gains strength. Moreover, maturation of the industry worldwide will, no doubt, be accompanied by the growth in protectionist policies already seen in the pharmaceutical industry--in particular, U.S. companies will have problems with standards, testing, and certification procedures in Japan and Western Europe. Neither quotas, tariffs, nor government procurement is likely to play an important role in this regard.

The two case studies that follow contain some of these elements and offer an indication of how some companies have dealt with these issues in the course of business. Table 4.1 summarizes the strategies that have been adopted by these firms and the circumstances that have shaped these decisions.

TABLE 4.1
Summary of biotechnology case studies

Firm No.	Total Assets ($ Mil)	Strategy	Reason
1	55	Joint venture in Japan with nonpharmaceutical company; limited technology transfer	Need for non-U.S. source of manufacture to circumvent U.S. regulations. Need for local partner to deal with Japanese regulatory authorities and system need for capital; desire to retain rights to U.S. market
2	44	Sells through Japanese trading company; technology transfer outside of company's area of proprietary research; in Europe has own marketing representative	Perceived difficulty of marketing in Japan due to cultural differences and complex regulatory structure; need for capital, but desire to maintain an area of proprietary research

CASE NO. 1

The company was organized in April 1980 with the aim to develop, produce, and market commercially useful products using genetically engineered microorganisms created by means of synthetic DNA and recombinant DNA technology. The company has technical expertise in biochemistry, protein chemistry, cell biology, microbiology, fermentation, protein purification, pharmacology, immunology, organic chemistry, and process development. The firm's approach to product development emphasizes second-generation technology, that is, the cloning and expression of synthetic, instead of natural, genes to produce products with higher efficacy, better yields, and reduced manufacturing costs. The company strives to develop products for high-volume applications that can benefit from the special technologies proprietary to the company. The markets served include human therapeutics with vaccines, antivirals, and growth regulators; human diagnostics with hybridization probes and immunoassays; animal health care with growth regulators and other therapeutics; and specialty chemicals with products produced by hydrocarbon oxidation. Although the company now has five products with high market potential, it has not yet brought one to market. As a medium- to long-term goal, the company hopes to develop into a full-fledged pharmaceutical company.

One of the company's most promising products is in the area of human therapeutics. Recently, the company's advanced microsequencing and other proprietary technology led to the isolation, cloning, and expression of a human hormone that controls the production of red blood cells. Initially, this hormone is expected to be used to treat anemia, a condition in which the body's production of the hormone is often insufficient to sustain red blood cell production. At this time there are no other related alternative therapies. Repeated transfusions, the treatment used currently, are not only costly but also expose the patients to the risk of contracting hepatitis, acquired immune deficiency syndrome (AIDS), or other diseases. Other possible applications include the treatment of certain anemias caused by chemotherapy. The company has applied for patents in those countries with potential markets for the product. The potential world market for this product is estimated at $100-$200 million. In the United States alone, there currently are about 1,200 dialysis centers; the U.S. market size is estimated at $40-$60 million.

These major breakthroughs in the production of this hormone have been made during the last year-and-one-half. When the discovery was made, the company made a formal announcement soon after because it needed to raise additional capital. The announcement was read by a Japanese malt beverage manufacturer that was one of about six companies unsuccessfully working to develop this same hormone; the Japanese company contacted the company immediately and arranged for a meeting within the week. Within three months, the two companies had made a formal agreement to form a joint venture with respect to this one product. Ownership of the joint venture is divided evenly between the two parties. The company contributed $4 million in cash and proprietary technology, and the Japanese partner contributed $12 million in cash. The agreement provided that the company would own the U.S. rights, its partner would own the Japanese rights, and the joint venture would determine how to do profit sharing in the rest of the world. The company has three persons on the board and maintains control by having its own chief executive officer (CEO) function as CEO of the joint venture. The agreement allows for a totally open exchange of information with respect to this one product. At present, all the research being conducted on the hormone is being paid for by the joint venture. The research has been progressing as planned and on schedule with established milestones.

From the beginning, the company was particularly attracted to the idea of a Japanese partner for several reasons. Having a source of manufacture in Japan allows the company to circumvent some of the long and arduous FDA certification processes that are required if the company depends on U.S. production of the hormone, but are unnecessary if the company is selling out of Japan. Basically, the time and money involved in the registration process in the United States make it less attractive for companies to manufacture in the United States than abroad. Also, the Japanese partner is in a good position to act as a liaison between the company and the Japanese food and drug regulating authorities. An influential Japanese partner can be an important entree to the Japanese market since U.S. regulations stipulate that in order to export a drug that has not yet received FDA approval, a formal request to import that drug must be made by the Japanese government. Further, the Japanese partner was in a position to infuse large amounts of needed cash into the development of the product. Most importantly, the Japanese partner was willing to give capital in exchange for the Japanese rights and a negotiated share of the rest of the world, excluding the United States. Hence, the company was able to negotiate a deal in which it was able to retain U.S. rights and have a relationship with a firm that could supply a fairly high inflow of cash and serve as an overseas manufacturer. This particular hormone also is a product that lends itself well to an arrangement such as this because it can be marketed directly to the 1,200 or so dialysis centers in the

United States and, therefore, does not require a U.S. pharmaceutical company to facilitate sales and distribution. In the view of the company, these factors were the prime motivation for transferring its proprietary technology.

The company believed the only way it could preserve rights to the U.S. market for its product was to form a venture with a non-U.S. firm. In the past, the company has tried to negotiate deals with U.S. pharmaceutical companies in which the company tried to offer European rights in exchange for retention of U.S. rights. This approach has not been successful. For the most part, U.S. pharmaceutical companies have made it difficult for biotechnology to attach a high negotiating premium to their proprietary techniques and processes, insisting that the same technology can be obtained through another source. U.S. pharmaceutical companies want control of the U.S. market and do not want to pay high prices. As a result, the company feels that unless the U.S. pharmaceutical companies' marketing force is required, it is more attractive to look outside the United States for needed capital. In the company's opinion, it receives basically the same treatment from foreign pharmaceutical companies as it does from U.S. pharmaceutical companies with one very significant difference: it is able to retain U.S. rights when it enters into a deal with a foreign pharmaceutical firm.

There are formal limitations in the agreement with respect to the nature of the technology the company is required to transfer to its Japanese partner. The company is obligated to transfer only preferred systems for producing this specific hormone. The systems the company has transferred to its partner, though highly proprietary in nature, are incapable of easy modification to produce other substances; they are not universal vector systems. As a result, the company is confident it will be able to carefully control its technology, despite the required transfer. Furthermore, the Japanese partner is not involved in any competing business with the company, though the two may have areas of mutual interest in the future.

Of the estimated 45,000 patients currently on dialysis in Japan, approximately 27,000 would benefit from this hormone treatment. Using these estimations, the potential market in Japan is about $20 million, which is about one-third to one-half the estimated U.S. market. The companies believe that the European market for the product will be equal to, if not larger than, the potential U.S. market. For the sake of analysis, if it is assumed that the U.S. market is $40-$60 million, the Japanese market is $20 million, and the European market is $40-$60 million, and the company retains the U.S. market and one-half of the European market, the company still has control of 60-64 percent of the worldwide market for the product, worth roughly $60-$90 million.

The joint-venture agreement has been in force since May 1984, and the company has been very pleased with the relationship with its Japanese partner. Thus far, the two companies have been working very well together, and the project has been proceeding on schedule and as originally envisioned. The company hopes that if all continues to go well, the two companies can expand their agreement in the future to include cooperation in other areas of mutual interest.

CASE NO. 2

The company is a publicly held firm in the biotechnology industry. The company focuses its proprietary research on the production of enzymes and specialty chemicals. The company incurred a net loss in 1982, 1983, and 1984. These losses reflect a large commitment to proprietary research projects and product development activities.

Soon after establishment, the company began selling its services abroad. Historically, the company's contract R&D business has been in the area of pharmaceutical products. Initially, the company sold through an American firm that represented it in Europe and Japan. This representative, however, proved to be unsuccessful. After a year or so, the company decided to establish a major Japanese trading company as its Japanese marketing agent. For the European market, the company decided to base one of its own marketing representatives in Europe full-time. The company's contract research revenues have come in approximately equal proportions from the United States, Japan, and Western Europe.

Over the last six years, the company has had a number of R&D contracts in West Germany, Sweden, Switzerland, and Italy. The company has not had any major research contracts in France. Many of the large French companies considered to be potential customers for the company are either wholly or partially state-owned. As nationalized companies, these firms are not going to spend their research money overseas. Exchange rates have also been working against the company in Europe. Also, several of the European countries had a good base in genetic engineering research and as a result were not good potential customers for the company's training programs.

The company received its first R&D contract in Japan. Most of the work the company has done for the Japanese has been in genetic engineering: molecular genetics, microbiology, biochemistry, and, to some extent, fermentation. The majority of these contracts involve only the construction of a microorganism with certain specified properties; only one contract involved purification of material for preclinical trials. The company has had only one R&D contract in Japan involving the technology that it is specializing in on a proprietary basis--namely, specialty chemicals and enzymes.

The company also has been involved in substantial technology transfer projects through interaction with foreign scientists at its U.S. facilities. Several of these interactions have been substantial and intensive.

For the most part, the company negotiates R&D contracts worldwide on a "fee-for-service" basis. These contracts generally have a bonus provision in the event that the company does a spectacularly good job, resulting in a very high rate of production. Additionally, these contracts have royalty provisions based on net sales. In the majority of cases, the company retains ownership of the technology, and licenses the party contracting its services for specific applications of the technology. Generally, these agreements are for worldwide rights to the technology for a narrowly defined product application, though some agreements do not give the licensee worldwide rights.

Thus far, none of the contract R&D performed by the company has led to the commercialization of products on the Japanese market for the Japanese client. A few products are currently going through the regulatory cycle. One of these products, which was developed for a Japanese firm, is going through the regulatory process at the same time as a competing product that was developed simultaneously by a group of Japanese firms. Although it is certain that one of these products will be commercialized in Japan, there currently is a complicated debate over the patent positions of the two groups.

The company attributes its early success in the Japanese market to two main factors: the relative superiority of U.S. technology and restrictive research guidelines in Japan. Five years ago, Japanese companies had not developed significant in-house research capabilities, and U.S. research technique was considered state of the art in many areas of genetic engineering. Also, the Japanese regulations governing the types of research in genetic engineering that could be performed were very restrictive; by contracting U.S. firms to conduct this research, Japanese companies were able to expedite their research needs. In the early days, Japanese firms preferred to contract from the United States to gain needed technology and a head start over firms operating domestically. Japanese guidelines have now eased up with the result that a lot of companies are performing basic genetic engineering themselves. As a result, the company is experiencing a drop in demand for these services.

Each time the company receives a contract from a Japanese company, it must be approved by the Japanese government. The contract is not official until the government has put its seal of approval on it. Thus far, the company has not encountered any difficulties receiving quick approvals. The Japanese government, however, does support a lot of basic research in an area in which the company is particularly strong, and the company is beginning to see some difficulty in selling that technology in Japan. Japanese firms are not going to buy from the company technology that they can get from the Japanese government.

The company was recently approached by a large Japanese company to work on a project that involved biotechnology. This Japanese firm had also contacted several other U.S. firms; the entire project would cost multimillions of dollars and was to be divided into parts or sections. It was evident to the company that this was a public relations effort in the sense that the Japanese firm was committed to spending a sizeable amount of money in the United States in a very visible way. Since this is a relatively new area of technology and there were good financial incentives, the company was initially interested. Unfortunately, the Japanese firm wanted to structure the deal in such a way that it would own all the resulting technology. The company could not possibly agree to the terms. Indirectly, it sent a message back to Japan that if the Japanese firm is trying to develop a favorable view of Japanese willingness to spend money on R&D in the United States and have a positive public relations impact, the Japanese company better modify the terms of the deal; otherwise it will have just the opposite of the desired effect. This proposed project has gone no further.

The company has been relatively satisfied with its relationship with its Japanese trading company. The trading company takes commissions on the value of the research contract revenues in any given year. Seeking the representation of a Japanese trading company was the only option the company seriously considered. Given the inherent complicated regulatory processes involved and the cost of maintaining a permanent office in Japan, the company believed that using a trading company was the only choice. Even if the company were to grow substantially, it would still use a trading company because of the perceived difficulty of marketing in a country whose business culture is so different from that of the United States.

The company, however, did cite some inherent problems with such a marketing relationship. Since all large Japanese trading companies have their own family of related Japanese firms, the company tends to be introduced only to a particular subsection of potential Japanese customers. The company also believes that trading companies tend to be more enthusiastic about working hard when business is booming and prospects look good. Biotechnology, in general, is a tougher market to compete in today because many of the firms that were contracting the company's services three or four years ago now have significant internal capabilities to perform R&D. The company believes that this has affected not only the trading company's behavior, but also the underlying attitudes of the people who are on the company's account at the firm. The company also stressed that it is very difficult to change one's trading company in Japan once a relationship has been established.

The company has no plans to manufacture in Japan in the future. In terms of the company's main product, the current structure of the world market is such that Japanese manufacturing makes no sense. Manufacturing that product would require a very large plant to benefit from economies of scale, and shipping costs from Japan would be prohibitive. In terms of the company's other products, some of which are currently being evaluated by several Japanese companies, the company will probably opt to license the technology to Japanese firms to manufacture those products and supply some of the key raw materials to those Japanese firms. The potential market, however, is not large enough to justify the establishment of a manufacturing facility in Japan.

NOTES

1. U.S., Congress, Office of Technology Assessment, Commercial Biotechnology: An International Analysis, OTA-BA-218 (Washington, D.C.: Government Printing Office, January 1984), pp. 92-93.

2. Jeff Trewhitt, "U.S. Midwest Makes Pitch for Biotech-Research Firms," Chemical Engineering, April 15, 1985, pp. 27-28.

3. James S. Howard, "Deep Pockets," D&B Reports (July/August 1984), p. 24.

4. Office of Technology Assessment, Commercial Biotechnology, pp. 67-71.

5. U.S., Department of Commerce, International Trade Administration, "High Technology Industries: Profiles and Outlooks, Biotechnology," Table 11 (draft), July 1984, pp. 137-146.

6. Bio/Technology, April 1984, p. 33.

7. Kim Alayne Walton, "Cloning Around, Dr. Panglass Is Alive and Well in Bio-Tech," Barrons (January 7, 1985), p. 14.

8. OTA, Commercial Biotechnology, p. 65.

9. U.S., Department of Commerce, International Trade Commission, International Developments in Biotechnology and Their Possible Impact on Certain Sectors of the U.S. Chemical Industry, USITC Pubn. No. 1589 (October 1984), pp. 26-29.

10. Department of Commerce, "High Technology Industries," pp. 119-122.

11. Ibid., p. 87.

12. OTA, Commercial Biotechnology, pp. 518-519.

13. Department of Commerce, "High Technology Industries," p. 89.

14. OTA, Commercial Biotechnology, p. 512.

15. Ibid., p. 511.

16. David E. Sanger, "Biotechnology's Patent War," New York Times, March 19, 1984, pp. D1, D5.

17. International Trade Commission, International Developments in Biotechnology, pp. 29-31.

5
COMPUTERS

Over the last thirty years, the U.S. computer industry has undergone extremely rapid technological change which has helped it to achieve a position of international leadership. The industry has long been one of the fastest-growing sectors of the U.S. economy. The only decline in the value of U.S. computer shipments was during two recessions in 1971 and 1975.[1] Since 1972, the overall rate of expansion has averaged 18.1 percent (compounded in current dollars), and despite the slight 1981-1982 decline to roughly 11.3 percent, most industry analysts expect a 15-17 percent annual growth rate to continue at least through 1990.[2]

The world's first commercially available large-scale electronic computer, Univac I, was produced in 1951 by the U.S. company, Remington-Rand. Other first-generation computers were soon available, including the Univac II and IBM's 701. Initially, the main customers were large U.S. corporations and U.S. government. In the late 1950s, transistorized second-generation computers appeared on the market in the United States. These computers included the IBM 7090 and 1401, RCA 300, Burroughs 3500, and Control Data Corporation CDC 1604. Second-generation computers are characterized by their use of transistors, advances in logic, and introduction of magnetic core memories.[3]

By 1960, six U.S. mainframe manufacturers had entered the industry, including Control Data, Burroughs, Honeywell, IBM, National Cash Register (NCR) and Sperry Univac. IBM emerged quickly as the industry's dominant firm. Third-generation computers were introduced in the mid-1960s, incorporating integrated circuit technology, larger and faster memories, modularity in design, and time-sharing capabilities.[4] Third-generation computers included the IBM 360 series, the RCA Spectra 70 series, the Burroughs 6500 system, and the CDC 6000 series.

The fourth-generation computer systems introduced in the 1970s featured large-scale integrated circuitry, modular memory components, standardized communications systems, networking, remote diagnostics, mass storage, data base orientation, and distributed processing. These new systems combined reduced cost of calculations with an increase in computing speed. This development spurred the onslaught of so-called IBM compatible machines because of the lower cost to entry; these manufacturers included Amdahl, Fujitsu, and Hitachi.

Since the early 1970s, minicomputers and microcomputers have become powerful enough to perform tasks that only large mainframe systems could handle earlier. As computer prices have fallen and business applications for software proliferated, low-end microcomputers have become one of the fastest-growing segments in the computer industry, with personal computer sales increasing from 77 million in 1976 to an estimated 4 billion in 1982.[5]

High growth rates in certain sectors of the U.S. computer industry have been highlighted by the emergence of small firms responsible for such technological innovations as minicomputers and microcomputers, and a wide variety of storage and terminal devices. This industry segment has been pioneered by start-ups such as Apple Computer and Commodore International. Since 1977, the computer industry has witnessed an unprecedented number of start-ups spurred by expanding demand, readily available venture capital, steady reduction of component cost and the development of standard software systems. Between 1976 and 1982, the number of firms in the computer industry (SIC 3573) increased 132 percent from 695 to 1,618; small businesses (under 500 employees) as a percentage of this total increased from 93.5 percent in 1976 to 95.4 percent in 1982, according to the Small Business Administration Data Base. In 1976, small businesses accounted for 3.3 percent of total sales and 3.7 percent of total employment in the industry; by 1982, these figures had increased to 5.4 percent and 5.7 percent respectively. In terms of productivity, small firms have been outproducing large firms; in 1982 average firm sales per employee in firms with under 500 employees was $75,000 versus $64,500 in firms with over 500 employees.[6]

Most innovations in the computer industry have originated in the United States. This is reflected in world market shares of U.S.-based computer firms. The share of world production held by the United States has risen from 55.4 percent in 1978 to 66.2 percent in 1982. U.S. export performance has remained relatively constant over time, increasing from 43 percent in the mid-1970s to 47.5 percent in 1981; this figure then dropped to 45.5 percent in 1982, in part because of the strong dollar and recession abroad.[7]

The French computer market is the fourth-largest in the world; in terms of hardware it was valued at $4.1 billion in 1983. The 1983 computer market in France was 70 percent mainframe computers, 20 percent minicomputers, and 10 percent micro-computers.[8] Total production for the computer sector was $3.87 billion in 1983 and estimated at $4.32 billion for 1984; 1987 projections are as high as $8.1 billion with a real growth rate of 17 percent between 1984 and 1987. Currently the two fastest-growing segments of the French EDP equipment market are minicomputers and microcomputers, which are expected to grow at an annual rate of 20 percent and 35 percent, respectively.[9]

The French computer industry is dominated by one national-ized supplier, Groupe-Bull, which manufactures all types of machines from large mainframes to mini- and micro-computers. Groupe-Bull is the largest European computer manufacturer, and represents more than a quarter of the French computer population in value. U.S. suppliers represent 70 percent of the French computer population with IBM constituting 43 percent. The mainframe and minicomputer markets, where IBM and DEC are strongest, constitute 65 percent of Bull's sales.

The beginning of France's extensive support of its domestic computer industry followed General Electric's (U.S.) 1964 pur-chase of a controlling interest in the leading French computer firm (Machine-Bull) and a French dependence on the United States for large-scale computers. In 1966, the French government (GOF) launched a comprehensive program, the Plan Calcul, to direct development of an independent French computer industry, and formed Compagnie Internationale pour l'Informatique (CII) as the principal French-owned computer manufacturer. For fifteen years, the GOF provided more than $1 billion in support to the French computer industry, chanelling most of the funds to CII from 1967 to 1975 and to its successor, CII-HB, from 1976 to 1982.

Plan Calcul was carried out in four stages. The first stage (1967-1971) emphasized the development of small- and medium-scale computers and peripherals. The financial support for CII during this five-year period was sizeable--$80 million for develop-ment projects and $8 million in low-interest, long-term loans, and $25 million in private financing guaranteed by the GOF. During this stage Systemes et Peripheriques Associes aux Calculateurs (SPERAC) received $15 million in R&D contracts over five years and $3 million in reimbursable loans. SPERAC was responsible for most of the peripheral development part of the plan. Institut de Recherche d'Informatique et d'Automatique (IRIA) was established to promote long-term computer research. Intertechnique was responsible for development of minicomputers.[10]

The second stage (1972-1975) experimented with the notion of a European solution for the computer industry. This stage included an agreement with the GOF that gave CII $140 million in the form of research grants. By 1972, CII had developed and produced four commercial computer systems, a minicomputer and two military computers. In 1973, CII joined with Philips and Siemens in a tri-national firm, Unidata. The Unidata agreement called for collaboration in technical development and marketing among Siemens (FRG), Philips (Netherlands), and CII. GOF expenditures on Unidata reached approximately $1.104 billion from 1973 through 1975. Unidata was later dissolved. Through procurement preference (CII was the preferred supplier at market prices), CII gained an increased share value of the installed computer base in the French government and private enterprise, rising from 15 percent in 1971 to 25 percent in 1975.[11]

During the third stage (1976-1978) the GOF merged CII and Honeywell-Bull to create CII-HB. Ownership in the new company was 53 percent French and 47 percent Honeywell Information Systems. The merger accords specified that CII-HB would receive $736 million in government contracts and $221 million in grants from 1977 to 1980 from the French government. In May 1977, the Components Plan was announced. In 1978, the GOF gave grants totalling $18 million each to Thomson-CSF and Radiotechnique Compelec, Philips' French subsidiary, for R&D on integrated circuits, $37 million to EFCIS for production of MOS circuits and $37 million to Eurotechnique for production of NMOS and CMOS circuits, over a five-year period.[12]

The fourth stage (1979-1982) introduced demand stimulation in 1979; the GOF launched a program to provide $460 million over a five-year period to expand the use of computers in education, industry, and small businesses; for continued aid to CII-HB and producers of small computers and peripherals; to aid in the development of integrated circuits; to support development of TRANSPAC and to fund the 1983 launch of the international telecommunication satellite Telecom I.[13]

The French government has two important aims in its strategy toward development of the domestic computer industry; one is to increase its domestic market and French computer usage, and the other is to increase the market share of its domestic computer manufacturing firms thus increasing sales volume and revenues. The GOF began programs to stimulate domestic demand in 1979 during the fourth stage of the Plan Calcul. Beginning in 1979, a five-year program to expand the use of computers in the French educational system, small businesses, and in French industry through increased promotion of CAD/CAM and automated production facilities was launched.

More recently, the World Center for Computation and Human Resources was formed by Raj Reddy. It was subsequently led by Jean-Jacques Servan-Schreiber, an outspoken supporter of computer literacy, and now has an annual budget of $15 million. The center runs computer workshops, funds software development, and promotes integration of computers into French society. The center, along with the Ministry of Education, was recently involved in a $156 million purchase of teaching computers for the French school system. Servan-Schreiber has since left the center after a dispute with the French government reportedly over whether French or U.S. computers would be used for the program.

Preferential government procurement is the most significant targeting measure used by the French. Since the early 1970s, France has had stringent bureaucratic procurement procedures in place to support Bull and other domestic suppliers. This preferential treatment meant these agencies and companies were to buy from CII whenever possible; as far as can be determined, these agencies and companies paid market prices for these goods. In 1973, for example, these purchases amounted to 80 percent of CII's annual sales that year. Today, preferential public procurement for national sources is epitomized by the French manufacturers' 63 percent share of the government's installed base (versus a private market share of only 45 percent). Preferential treatment continues; recently, after Solmer Steel Works placed an order with Univac, government officials sought and obtained cancellation of the order in favor of CII-HB equipment.

These targeting measures have had a twofold effect. First, they have succeeded in giving French companies a larger share of the market than they would have under free-market and openly competitive conditions. Second, these policies have encouraged non-French companies that wish to be successful in the market to contribute significantly to the French economy. Digital Equipment, for example, is spending FF200 million building a production facility at Valbonne in the south of France; Hewlett-Packard, along with Wang and Apple (which do not have manufacturing facilities in France), has announced it will purchase more French components to increase the French content of its products.

Despite these efforts, the French computer industry has not increased its competitiveness in world markets. The share of world computer production held by France has decreased substantially over the past years from a high of 13.2 percent in 1977 to 7.4 percent in 1982. French exports also have declined from a high of 12.9 percent in 1976 to 7.9 percent in 1982. The domestic market, however, has been growing. The import penetration ratio increased by 31 percent to 51.6 percent of apparent consumption in France in 1982.[14]

The German computer industry includes two of Europe's most competitive firms, Siemens and Nixdorf. In 1981, these two firms accounted for 21 percent and 22 percent of the German market, respectively.[15] Nixdorf, considered to be the most internationally competitive European computer firm, manufactures primarily small systems intended for business applications. Nixdorf has a global outlook; it currently generates almost half its sales in West Germany and 20 percent in the United States.

Though Germany is home to some of Western Europe's most competitive computer manufacturers, its relatively open market has allowed American firms to obtain a market share comparable to that common in the less-competitive but more-protectionist countries of Europe. The United States plays an important role in the German market through direct export, shipments from non-German subsidiaries, and the internal production of American firms. West Germany maintains a persistent deficit in computer trade, with nearly half its imports coming from U.S. subsidiaries and the rest of Europe.

The German government has actively sought to promote the development of its domestic computer industry. In 1977, the German government established a significant network of support programs, which covered twelve financial plans in the three main areas termed crucial to future computer development: industrial research and development, data processing applications, and manpower training. Between 1977 and 1980, funding for these programs reached approximately $650 million. During the 1980s, the German government is funnelling funds into research and development assistance to firms and universities through the Federal Research Ministry of Technology Center in Berlin. Funding for the center reached about $150 million in 1982.[16]

The German computer industry also has benefitted indirectly from software development programs. Beginning in 1980, the Federal Ministry for Research and Technology (BMFT) began funding a computer software program aimed at upgrading software technology research through the improvement of specifications, operating techniques, quality control, and servicing. The goal of this program is to lower production costs and improve reliability of German software products. This program is progressing in various stages: integration of software/hardware development in microcomputers, funded at DM4.5 million (1980-1983); improved product reliability, funded at DM2.2 million (1982-1983); protected data automation systems for separated offices, funded at DM5.4 million (1982-1984); and computer-aided software production, funded at DM3.5 million (1982-1983).[17]

There has been evidence that standards and equipment testing procedures have been used in the past to discriminate against U.S. products in West Germany. Computer manufacturers exporting to West Germany are required to meet West German standards on electromagnetic interference (EMI), which are developed by the German Society of Engineers (VDE). Further, it is required that the VDE perform certification tests--U.S. manufacturers must, therefore, incur costs of bringing VDE engineers to the United States to perform the same tests that are already performed by U.S. manufacturers. In addition, since West German product safety for data processing equipment is not harmonized with American UL or Canadian CSA standards, U.S. manufacturers have had to design their products using VDE-certified components.[18]

The United Kingdom is home to Europe's largest computer manufacturer, International Computer Ltd. (ICL). ICL has about a 30 percent share of the domestic computer market. Its main competitor in the domestic market is IBM, with a 50-60 percent share. Though growth in demand is expected in the U.K. market, much of this increased activity is expected to be absorbed by U.S. competitors and Japanese firms, through OEM agreements.[19]

Like its European counterparts, the government of the United Kingdom also has been active in promoting the development of its computer industry, with the use of grants, subsidies, loans, and publicly supported research. The Department of Industry has committed 80 million pounds for the development of information technology developments. The program has three distinct parts: the Product and Process Development Scheme, the Electronics and Avionics Requirement Board, and the Software Products Scheme of the National Computing Center. The government directly supports public research activity and furnishes grants to both firms and users to encourage more extensive computer application and usage. In addition, the government pledged to make selective purchases of automated office equipment to establish pilot studies and demonstration programs of the equipment within the government.

The Department of Industry also is involved in a joint research program, the Alvey program, which is aimed at development of advanced technologies for fifth-generation computers. This five-year, 350-million-pound effort covers the following topics: software engineering,   intelligent knowledge-based systems, interface beween humans and machines, and very large scale integration of electronic components. The program's current project is known officially as Flagship. Flagship, based on work at Imperial College and at London and Manchester universities, will run declarative fifth-generation computer languages.

This $20 million, three-year project is the largest single effort funded under the Alvey program. ICL will lead the project team, which also includes Plessey. Plessey is responsible for creating the first large-scale demonstrator.[20]

ICL and other British computer manufacturers have benefitted from preferential government procurement over the years as a by-product of these government targeting measures. As a result, U.S. companies have had trouble competing against these firms in their domestic markets. Additionally, the British government has repeatedly provided emergency financial assistance to ICL during its periods of severe financial strain; between 1981 and 1983, the government lent ICL over 400 million pounds through a series of loans. Moreover, the British government also has blocked the purchase of ICL by foreign corporations on several occasions.

The European computer industry also receives assistance through ESPRIT, which is currently funding a collaborative project by four British and three French groups aimed to cut the entry cost into supercomputing by adopting a parallel-processing approach instead of building a specialized machine. ESPRIT is providing $6 million for the effort, which will be based on a large array of transputers, the Inmos International microprocessor designed to support parallel processing. The project team, which will be guided by Southampton University, includes Royal Signals and Radar Establishment, Inmos International, Thorn-EMI, the University of Grenoble, Telmat, and Apsis.[21]

Japanese planners view the computer as playing a pivotal role in the nation's planned transition to a "knowledge-intensive" information society. Computers are viewed as a high value-added product with a strong growth potential, which also are capable of improving productivity in a wide range of industries. Thus, computer hardware and software development has received high priority in the Japanese government's industrial policy as well as in private-sector research efforts.

The Japanese computer industry has grown rapidly in recent years (17.5 percent annual rate compounded) and is now the second largest, behind that of the United States, in the world. In 1982, exports soared to 340 billion yen, 19.6 percent of production, up 32 percent over 1981, when 13.2 percent of total production was exported. The value of computer imports, while still rising, is growing at a considerably slower pace. Until 1981, Japan was a net importer of computer technology. However, impressive recent advances in technology seem to ensure a strong long-term export growth potential.

The Japanese computer market as a whole can be characterized as highly competitive, with aggressive price cutting frequently used to gain market share and/or to retain key customers. Nevertheless, domestic firms are frequently favored in procurement, particularly in the case of sales to the government, NTT, or firms belonging to the same industrial group as the seller. Fujitsu, IBM-Japan, NEC and Hitachi each provide a broad line of computer systems and peripherals, including mainframes. Also providing a broad computer product line, but less successful in terms of market share, are Mitsubishi Electric, Nippon Univac, Burroughs, and NCR-Japan. Toshiba and Oki Electric dropped out of the mainframe business some years ago and now focus on smaller computer systems, peripherals, and process control systems. They are joined by a host of other firms from the consumer electronics and office machines industries including Matsushita, Sharp, Sony, Sord, Casio, Uchida Yoko, Sanyo, and Ricoh. Most of this latter group are focusing on the burgeoning field of office automation along with peripherals.

The structure of the major Japanese computer firms, and of Japanese industry as a whole, aids in their competitiveness. They are, in general, highly vertically integrated. In addition, many of them maintain close links with large industrial groups centered around commercial banks, thus providing them with a large captive market as well as ready access to funds. NEC, for example, maintains a very high debt-to-equity ratio that would be unacceptable in the United States but is not a problem in Japan, in part because of its close relationship with the Sumitomo Bank/Sumitomo Group. The firms also enjoy the opportunity to participate in GOJ/NTT-sponsored R&D projects, as discussed below. In addition, NEC, Fujitsu, Oki, and Hitachi benefit from their position as the major equipment suppliers to NTT. Besides gaining lucrative sales, they perform contracted research and/or joint research with NTT and are frequently the recipients of high-level technology transfer from that organization. Many of the larger Japanese computer companies also benefit from horizontal integration with related sectors such as communications and robotics.

Since the early 1960s, the Japanese government has been extensively involved in supporting computer R&D and promoting greater domestic usage of computer systems. Targeting has taken a variety of forms, evolving as the industry matured and attained international competitiveness. Tariffs and quotas played an important role in protecting the industry until about 1975, but have become unimportant since. R&D support has been an important and continuing policy instrument, evolving from a pattern of targeting successive generations of IBM mainframes to the current emphasis on the development of advanced technology on the frontier of current knowledge. The major computer-related policies and projects are described briefly below.

- Fifth-generation computer. An ambitious long-term R&D effort to create a next-generation computer incorporating artificial intelligence.
- High-speed science and technology computer. R&D on parallel processing architecture and high-speed memory devices with the goal of creating a supercomputer with a speed of 10 gigaflops by 1989.
- Fourth-generation computer. A two-phase project aimed at developing the basic hardware (VLSI), operating system software, and peripheral equipment required to compete with future IBM computers.
- Optical measurement and control system. The goal of this project is the creation of a complete system for control of large-scale industrial processes through the use of optical elements to sense and transmit data.
- NTT-DIPS series computers. Joint R&D between NTT and the major computer manufacturers to produce a series of advanced mainframes.
- Additional accelerated depreciation and tax deductions are provided for computer purchasers and producers.
- Low-interest government loans are provided to the Japan Electronic Computer Company (JECC), a consortium of producers, to facilitate computer leasing.

The most significant nontariff barrier at present appears to be in the area of procurement. The Japanese national and local governments and associated organizations purchase a significantly higher share of domestic computers than do firms in most industrial sectors. Formerly important computer sector NTBs such as quotas and foreign investment restrictions are no longer in existence. Access of foreign firms to government-held computer-related patents has apparently been liberalized with the recent agreement between MITI and IBM to allow IBM to license all such patents held by the Agency of Industrial Science and Technology. The GOJ also recently announced that it will abolish its now low computer tariffs in the near future.

Extensive targeting of domestic computer industries in Japan and Western Europe has resulted in the formation of policies that discriminate against nonlocal products and the levying of performance requirements that force nonlocal firms to alter the way they do business in these countries. Preferential government procurement has posed a significant impediment to U.S. companies trying to sell computers in France. In Case No. 3, after the company had been publicly awarded a large contract by a nationalized manufacturer to install a large-scale computer system, government pressure forced the nationalized manufacturer to revoke the contract and give the award to Bull, the state-owned computer manufacturer. The company has repeatedly encountered overt preference for Bull's machines; as a result, the company no longer quotes the French central government as a matter of policy.

Incidents of discrimination against U.S. companies have gained significant notoriety, such as the recent case involving the $156 million purchase of teaching computers (120,000 micro-computers of varying brands) for the French school system. This order is the largest single purchase of computers in Europe, and part of an ambitious campaign to promote French computer literacy. Apple made an aggressive bid to win part of the sales order that involved the possible production of the Macintosh computer in France through a licensing or joint-venture agreement. Apple's overtures, however, were countered by open hostility from French computer manufacturers, such as Thomson SA and Groupe-Bull. As a result, French officials openly stated that the chief beneficiaries of the government computer literacy program will be French computer manufacturers.[22]

U.S. computer manufacturers also are encountering increased pressure to perform local R&D employing local workers, joint R&D with local companies, and to source their product locally. This is particularly strong in France where several U.S. computer manufacturers have already set up manufacturing facilities. The experience of these manufacturers is typified by the company in Case No. 4, which has slowly moved more and more of its production and R&D functions overseas to counter these mounting pressures.

In addition to preferential government procurement and performance requirements, companies have also encountered difficulty obtaining patents in Japan (Case No. 5). The three case studies that follow are summarized below in Table 5.1.

CASE NO. 3

The firm is a publicly traded company that specializes in the manufacturing, sale, and servicing of a wide range of computer systems and the development of related software and communications systems. The company offers a personal desktop computer in seven models, which takes advantage of the vast array of industry-compatible software. The company's personal computer connects to its own mainframe models as well as IBM mainframes. The company offers a wide variety of office systems, as well as large-scale systems that are among the largest general purpose computers offered by anyone in the world. Computer revenues

TABLE 5.1
Summary of computer case studies

Firm No.	Total Assets ($ Mil)	Strategy	Reason
3	over 500	Wholly owned marketing and service organizations throughout Europe with a manufacturing facility in Germany; two joint ventures with local Japanese companies; greater emphasis on developing products specifically for the Japanese market in the future	Nationalistic attitudes throughout Europe; "buy national" in France; difficulty selling to PTTs in Europe; teamed with local Japanese companies that were not competitors; heavy competition in large-scale computers from local Japanese companies and appearance of regulation in that market
4	502	In Europe and Japan, began with distributors and then set up wholly owned subsidiaries for sales and marketing; also wholly owned manufacturing subsidiaries in Europe	Increasing pressure to contribute to local economy; pressures to perform local R&D, joint R&D with local firm, or source from local companies; difficulty retrieving some of the costs associated with adapting the product to Japanese market

for 1984 were over $4 billion. The value of installed base, at $17 billion, is the second largest in the world computer marketplace. The company's marketing strategy emphasizes systems solutions involving unique combinations of software and hardware that are designed to address the specific needs of its customers. The company has a long history of involvement in the international marketplace.

The company involvement in the European marketplace dates back to roughly the early 1950s. The company established and maintains a wholly owned marketing and service organization in each country where it generates significant business; these include Spain, Holland, Great Britain, Italy, Switzerland, and West Germany. These markets are supplied about 80 percent from a U.S. source of manufacture and 20 percent from a manufacturing facility in West Germany. Today, the company generates approximately 50 percent of its total revenues outside the United States. Of that 50 percent, roughly 75 percent is generated in Europe. At 37.5 percent of total worldwide sales, the European market is a significant share of the company's total business.

The company's major customers in Europe are, in order of relative importance, the banking and financial community, public-sector manufacturing companies, and the transportation industry.

Although the company supplies many local postal authorities in Europe with general-purpose processing systems and some message-switching computers, the company does not manufacture central office or PBX equipment. Its total market share with European PTTs is therefore limited by this restricted product offering. Throughout Europe, the company's main competitor is IBM. In addition, there are local companies within each country market that compete directly with the company. Examples are: Olivetti in Italy, Siemens in West Germany, and Machine-Bull in France. These companies do, in the company's opinion, benefit to varying degrees from the preference shown to locally made products within each country. Although the company does not believe that having a manufacturing facility in Germany helps it to sell into other countries in Europe, it does believe that it helps the company somewhat in Germany. Nonetheless, the company has found that, in Germany, Siemens has a much larger share of the market.

The company has encountered exceptional difficulties competing against Machine-Bull in France. In the early 1980s, the company was awarded a contract worth roughly $8 million to install a large-scale multipurpose computer system in France. The customer was a state-owned manufacturing company. The contract was then later retracted and awarded to Bull. It had been rumored at the time that pressure had been exerted from high levels within the French government. The company threatened to go to court and was awarded damages as a result. As a matter of policy, the company normally refuses to quote to new customers of the French central government. The company's market share in France has remained stable over time. Sales in France have been fairly constant at approximately 10 percent of total European sales; this is inconsistent with the relative size of the French computer market. In addition, the company has been unable to benefit from the growth experienced by the French computer industry over the last few years.

The company believes that future growth in its European sales will take place in those countries where the company is already strong. Two such countries are Italy and Switzerland. Although Olivetti is very strong within Italy, its traditional strength has been in smaller-scale computers. Thus, the company's large-scale computer line has fared rather well in Italy. The company does not expect to increase its market share in France in the near-term future. Regarding the company's medium- to long-term prospects in Europe, the main threat to market share is perceived to be the Japanese move into the European computer market. The company is particularly alarmed by the recent takeover of a Spanish hardware manufacturer by

Fujitsu. This acquisition was made one month after the Common Market decision to allow Spain to enter the market in 1987. Since Spain is obviously going to be a low-cost source of manufacture within the market, this could give the Japanese a significant advantage.

The company has been involved in the Japanese market since the 1950s. In the early 1950s, it set up a joint venture in Japan with a Japanese partner. The joint venture is owned approximately one-third by the company, one-third by the Japanese company, and one-third is traded on the Tokyo Stock Exchange. This joint venture is solely a marketing and service organization; it currently employs 4,500 people and generates $500 million in revenues. The joint venture maintains a very Japanese image; of its 4,500 employees, only four are non-Japanese.

The company chose its partner on the basis that its name was very well known in Japan; it had an intimate knowledge of Japan and established contacts with every major Japanese company and thousands of smaller ones. The Japanese partner is not a potential competitor; it is strictly a Japanese trading company, and as such has had no conflicts of interest with the company. The company has been very satisfied with this relationship, which has now lasted over thirty years, and with its penetration of the Japanese market. The company believes that the influence and advice of a local partner has been invaluable, and attributes the company's success in Japan to the close cooperation between the two companies over the years.

Formal restrictions on small- and medium-scale computer imports led the company to enter into another joint venture in Japan in the mid-1960s. This joint venture manufactures only medium-scale and smaller-scale computers; the company continues to manufacture its line of large-scale computers in the United States. The Japanese partner, an electronics company involved primarily in the communications business, owns 30 percent of the joint venture; the company owns 45 percent; and the remainder is owned by several other Japanese companies. The joint venture has a very strong legal basis for the transfer of technology, which includes very specific licensing agreements, royalty agreements, and protection. The agreement also covers the rights to new products developed by the joint venture that are based on the company's technology. The company believes that this clear, written, and agreed-upon legal basis along with the good, solid relationship based on mutual trust that exists between the two business partners has contributed greatly to the stability and success of the joint venture throughout the last twenty years. This joint venture now employs roughly 650 people and generates $100 million in revenues.

The company has been relatively pleased with its progress in the Japanese market. Although the company's market share in small- and medium-scale computers has not been that great in Japan, it is basically in line with the company's market shares for the same products around the world. This reflects the fact that the company's strength has historically been in the large-scale computer market.

The company believes that the large-scale computer market in Japan has been regulated. Market share for the Japanese has been constant over the last few years at about 50 percent. The remaining 50 percent is left to be fought over by non-Japanese companies. Although the company could not quote its exact market share, it does have a major installed base and significant sales of $500 million a year. Its market share is, however, gradually shrinking. The company believes that this gradual decline most likely reflects a gradual strengthening in the position of the Japanese firms in the market.

The company believes it has not suffered as a by-product of the Japanese government's aggressive targeting policies toward the computer industry. The company attributes this to two main factors: (1) it moved into the Japanese computer market very early, during the 1950s, before the Japanese government put such a great emphasis on the development of the computer industry; and (2) it has maintained a profitable joint venture in Japan that has favorably impressed MITI.

To strengthen and maintain its market position in Japan, the company is now shifting its strategy to emphasize the development of products specifically for the Japanese market. The company recognizes that U.S. companies can no longer ship non-Kanji-speaking products to Japan, or products that are designed for Roman characters and retrofitted for Kanji characters. The company's competitors, such as Fujitsu, are addressing the needs of the market on the front end. As a result, these Japanese companies are developing products that are quicker and less expensive. The company is optimistic that with research and development efforts targeted specifically at the Japanese market, it will be able to reverse the slightly downward trend in its market share.

CASE NO. 4

The company is a thirteen-year-old publicly held firm in the computer industry specializing in providing computer systems to the engineering and scientific manufacturing markets. The company's three principal markets are engineering and scientific, commercial distributed data processing, and computer-aided design/computer-aided manufacturing (CAD/CAM). In 1984, the company had nearly $650 million in revenues; 42 percent of total revenues were generated overseas. The company's products are sold in forty-six countries around the world.

The company is well positioned for success in the CAD/CAM marketplace. Unlike most CAD/CAM vendors who offer only CAD/CAM products, the company has a full range of office automation, data management, and communications capabilities that complement its extensive CAD/CAM offerings in providing customers with truly integrated solutions. The company is one of the ten largest suppliers of CAD/CAM systems. CAD/CAM accounts for 15 percent of the company's revenues and represents the fastest-growing segment of the company's business.

Today, the company has twelve wholly owned subsidiaries and thirty-two international distributors of its products throughout the world. In Europe, the company has wholly owned marketing, sales, and service organizations in the United Kingdom, France, Germany, Switzerland, Italy, Norway, Sweden, Belgium and the Netherlands. Nearly two-thirds of the company's overseas revenues come from Europe. The company began selling in Europe through distributors in 1972 and then slowly began to buy out or convert its distributors to wholly owned subsidiaries. These decisions were based on market potential, the requirements of local customers and the political/economic environment in the country. Some markets simply grew too big for the distributor. The company began manufacturing in Ireland in 1976. This was motivated primarily by attractive tax incentives and the need to more closely support the growing European market.

The company began selling in Japan in 1978 through a distributor. In September 1984, the company established a wholly owned sales and service organization in Japan. The company will generate roughly 10 percent of its business there this year. To strengthen its overall position in the Far East, the company established similar subsidiaries in Hong Kong and Singapore within the last two years. The Far East is the company's fastest-growing region in terms of sales.

The company has clearly pursued a very aggressive policy toward penetration of overseas markets. The company believes strongly in the need for a direct presence overseas to best satisfy the customer's requirements. The company has built its reputation on its quality products and high-level of professional service. Moreover, the company has been willing to take the risks associated with hiring a large sales force prior to realizing the resulting revenue. Thus far, the company's strategy has paid off handsomely.

The company expressed some concern, however, about the increasing trend among central governments toward requiring a firm to contribute to the local economy either by manufacturing locally, by performing joint research and development with a local firm, or by sourcing local product. Currently, the company has official remote R&D sites in Ireland, England, and Australia. Official remote R&D accounts for a little over 10 percent of the company's total R&D expenditures. This figure does not take into account remote R&D for localization of product or specific product adaptation to market work that is often performed by local subsidiaries.

One of the company's most demanding markets is France. The French have very strict requirements for localization of product. Everything that appears on the screen must be in French, as well as user manuals and associated literature. The company had to spend substantial sums of money to localize the product. Most of this work had to be done in-country because U.S. translators seldom produce adequately translated products for the native user. Nonetheless, the French government does not consider this type of work as part of the company's necessary contribution to the local economy.

The company previously was able to satisfy its requirements by having a manufacturing facility in Ireland, which is part of the Common Market. However, the company has witnessed a fractionalization of the European market characterized by a strengthening of nationalism throughout Europe. As a result, the company is now unable to satisfy local requirements in the French or German markets by operating a manufacturing site in the Common Market.

Although the company has received a number of overtures from local governments in the past, it has not, to date, been involved in any substantial joint R&D efforts in Europe. Thus far, the company has been able to satisfy these requirements by sourcing local equipment. The company prefers this method, as long as it is not forced to source competitive equipment. The company did encounter such an experience recently in Australia, where it had to procure a substantial amount of equipment that directly competed with one of its own product lines from a local vendor. The company is, however, currently considering two such projects: one in the United Kingdom and one in France. At present the company performs no official R&D in France, but expects to have an official R&D site there by 1988 due to pressure from the French government.

The company's experience shows that the French tend to be less subtle about these requirements. If French authorities believe that a firm is acquiring a critical mass within France, but feel it is not contributing sufficiently to the local area, the authorities will rezone that firm to a less-desirable area that could be unsupportable based on the size of a company's operation, or from which it is logistically unfeasible to run a national sales effort.

The company is finding that the European countries, specifically France and Germany, are becoming more sophisticated in their requirements governing the contributions of foreign firms to the local economies. Initially, the firm began by distributing its manufacturing to local firms because this is easier than distributing development work. These countries have begun to realize, however, that although this is increasing local employment, it does not increase the country's technology base. These countries are becoming more specific in their requirements; they will ask a firm to perform joint hardware and software R&D with local firms and, in some cases, will even specify the type of hardware or software development that should be done. VLSI is an area in which specific interest has been shown.

The company believes that these demands, which it first encountered in Europe, are now spreading to underdeveloped countries. The company had been doing a good business in Mexico until the Mexican government expressed concern that the company was not performing enough manufacturing and/or R&D in Mexico. After that, the company's business nearly disappeared. The local authorities made clear to the company what it needed to do to rectify the situation. This same trend is apparent in the Indian market, which is just beginning to expand. Although the Indians have not be as blatant as the Mexicans, it is apparent from discussions between the company and Indian authorities that the Indian authorities expect some degree of technology transfer to take place.

These issues are somewhat different in the Japanese market. The company has to spend large sums of money to adapt its product for use in the local market because of the differences in the Latin and Kanji character sets. R&D which is performed for the Japanese marketplace is only applicable to that marketplace. The company still needs to make significant changes to this product for the Chinese market, for example. As a result, the company has trouble recouping these R&D expenditures in the marketplace. The company's Japanese competition obviously does not have this same problem. Modifications for Latin characters are applicable to the United Kingdom, Canada, Latin America, France, Germany, and other European markets. Thus, the Japanese competition has certain advantages in the internationalization of its product. Because of these differences, the company does not encounter the same pressures and obstacles in the Japanese market as it does in Europe and Latin America.

Nonetheless, the pressures encountered by the company around the globe are changing the way the company conducts its business. Increasingly, the company's manufacturing and R&D are moving overseas. The company does, however, see an organizational limit to the amount of R&D that can be conducted abroad. One of the main attributes of the company's product line is that it is highly integrated and its components are totally compatible. To achieve and maintain these attributes, the basic architecture of the system must be evolved at some central location. This becomes more difficult as the amount of remote R&D increases. There are, therefore, limits to the amount of remote R&D the company can perform.

CASE NO. 5

The company designs, manufactures, and markets a complete line of printers for use with microcomputers, minicomputers, and mainframe computer systems. The company's printers are designed primarily for business applications that require hard-copy multipart form output produced by impact printing. The company markets products worldwide to original equipment manufacturers (OEMs) and, through a network of distributors, to system integrators and end-users. Several printing technologies are employed in the product lines, including serial and line dot matrix printing for low to medium speeds, and band and chain-train technology for high-speed printing. The company's stock is publicly traded.

Early in its development, the company experienced very difficult times and was facing insolvency. Almost fortuitously, the company was approached at this time by a Japanese electronics firm that had just invested in a local (U.S.) disc drive manufacturer. The Japanese firm was interested in a similar venture with a much-needed capital, in return for a licensing agreement that gave the Japanese company the right to manufacture and sell the printers in Japan and the Far East. The company would have preferred not to enter into such an agreement, but urgently needed capital to continue operations, and, therefore, accepted the Japanese offer.

The Japanese were particularly interested in the line matrix printer because it had the potential to provide Kanji characters. The licensee, however, was not completely satisfied with the company's printer. In 1981, after some research and development work, which it performed using its own resources, the company modified the dot size, and this new printer sold very well in Japan. The Japanese licensee was quick to realize that this modified printer also provided distinct advantages over printers then being marketed in the United States for CAD/CAM and advanced graphics applications. At that time, the Japanese firm approached the company with a request to market and sell the new printer directly in the United States. While the company considered that this new printer utilized its patented technology, the Japanese company felt the design was sufficiently different and disagreed with the company's position.

After a discussion, the problem was resolved between the two parties by the company agreeing to market and sell the Japanese-designed printer in the United States on an exclusive basis. This printer has sold successfully in the United States, with fifty to one hundred units sold monthly. The company receives royalties on the products the Japanese licensee manufactures of the company's design, but does not receive royalties on the products of Japanese design. However, it is receiving a product from the licensee that it sells profitably in the United States. The result, then, of the original agreement between the company and the Japanese licensee has been that the company has lost the Japanese and Far Eastern markets for its own printer and has forgone the Far Eastern market for the printer designed by its Japanese partner specifically for Kanji, CAD/CAM, and advanced graphics applications.

The company realizes that these losses are due, in part, to its economic necessity at the time the agreement with the licensee was negotiated in 1975, and in part to its preoccupation with its efforts in the U.S. and European markets, and therefore not making a significant attempt to satisfy the specific needs of the Japanese and Far Eastern markets for a good Kanji printer.

Specifically, the company needed to provide a machine with a smaller dot size that could be adapted for Kanji characters. Since the company had already relinquished the marketing rights in that marketplace to its licensee, there were no incentives to develop or significantly modify its products for the market. Today, the company is isolated from the Japanese printer market, although it maintains a good relationship with its Japanese licensee.

From the beginning, the company had decided that acquiring patents for its design breakthrough in printer technology was a top priority. In 1975, the company applied for patents in the United States, the United Kingdom, France, Germany, Canada, and Japan. In the ten years that followed, there was no problem in obtaining patents in North America or Europe, but a patent for the design concept still had not been obtained in Japan. The company's patent case is currently being heard by the Japanese High Court. The chief reason given for denial of patent rights is the existence of prior art in Japan.

The company applied for a patent in the United States in August 1974, which was granted in March 1976. In August 1975, the identical patent was applied for in Japan. Despite the fact the claim in the Japanese patent application was repeatedly amended to add further limitations, the application was given a final rejection in April 1978. The company filed an appeal in July 1978, and in April 1979, received another Office Action that called for another amendment that further limited the scope of the patent claim. Despite the rather detailed claim of relatively limited scope then on file, the application was once again rejected in March 1984. On May 21, 1984, an appeal of the adverse decision was filed in the Tokyo High Court; final action is still pending.

The U.S. patent granted to the company in March 1976, affording reasonably broad coverage, is regarded in the industry as a pioneering patent application in the field of dot matrix line printers and has been at least partly responsible for the growth of the company from a small upstart to a company expected to have sales over $100 million in its dot matrix line printers in 1986.

In November 1982, the company became involved in Federal Court in a suit against a Japanese company that was selling a printer in the United States, which allegedly infringed upon the company's patents. The case was also accepted by the International Trade Commission in July 1983, and went to trial. During the trial, the Japanese company offered to settle the case, and the two companies reached an agreement that included a cash settlement and a provision that no new machines would enter the country after a certain date. After the settlement, both Hitachi and NEC abandoned their pursuit of the U.S. market in line matrix printers. If the company were able to obtain Japanese patents for its printer technology, the company could possibly bring proceedings against Hitachi and NEC, as well as other Japanese companies that are currently infringing upon its patents in Japan, which could possibly result in the payment of royalties to the company. Since the company's Japanese licensee would be covered by the company's Japanese patents, it could find itself in a more competitive position vis-a-vis the other Japanese firms involved in the printer market.

NOTES

1. The Futures Group, The Impact of Foreign Industrial Targeting on the U.S. Computer Industry, Report 701-138-03 (Glastonbury, Conn.: The Futures Group, April 30, 1985), p. 2.

2. Ibid., p. 58.

3. The Competitive Status of the U.S. Electronics Industry, prepared by the Electronics Panel, Committee on Technology and International Economic and Trade Issues of the Office of the Foreign Secretary, National Academy of Engineering; and the Commission on Engineering and Technical Systems, National Research Council (Washington, D.C.: National Academy Press, 1984), pp. 60-61.

4. Ibid.

5. Ibid., p. 62.

6. Small Business Administration Data Base.

7. The Futures Group, Impact of Foreign Industrial Targeting, p. 2.14.

8. U.S., Department of Commerce, CMP Industry Sector Analysis, by Myles Denney-Brown (Washington, D.C.: Government Printing Office, August 6, 1984).

9. U.S., Department of Commerce, International Trade Administration, International Market Research Profile on Mini and Micro Computer Systems in France (March 1984), pp. 3-12.

10. U.S., Department of Commerce, Bureau of Industrial Economics, French Government Assistance to Its Domestic Computer Industry: Lessons and Implications, no. BIE/IR-83-01 (June 1983), pp. 4-5.

11. Ibid., p. 6.

12. Ibid., pp. 6-8.

13. Ibid., pp. 8-9.

14. The Futures Group, Impact of Foreign Industrial Targeting, p. 2.14.

15. J. K. Paul, ed., High Technology International Trade and Competition (New Jersey: Noyes Publications, 1984), p. 71.

16. Ibid., p. 72.

17. The Futures Group, "Data Base on Industrial Targeting," 1984.

18. Ibid.

19. Paul, High Technology Trade, p. 73.

20. Kevin Smith, "Britain Makes Major Bid to Build Commercial Fifth-Generation Machine," Electronics, July 8, 1985, pp. 26-27.

21. "Electronics Newsletter," Electronics, August 19, 1985, p. 12.

22. "Apple Computers vs. French Chauvinism: Politics, Not Free Trade, Wins in the End," Wall Street Journal, March 1984.

6
SEMICONDUCTORS

There are currently more than one hundred American firms manufacturing semiconductors. The U.S. semiconductor industry is unique among the world's major producers in that it consists of both captive and merchant producers. Captive producers generally consume most or all of their own semiconductor output and may make additional purchases from the outside to satisfy internal needs. Merchant producers, though they may consume some of their semiconductor output, also sell their product to outside customers. Captive production accounts for approximately 30 percent of total U.S.-based semiconductor production. The value of captive IC production grew by 141.8 percent between 1978 and 1982. The value of merchant IC production, which accounts for 70 percent of total U.S.-based semiconductor production, grew by 99.2 percent. Captive and merchant production, however, are expected to grow at approximately the same rate through the mid-1980s.[1]

The field of captive semiconductor producers is dominated by IBM and Western Electric, which together accounted for 75.8 percent of IC production in 1982. IBM is not only the largest captive IC producer but also the largest producer of ICs in the world. In 1983, the estimated sales value of its output was $2.4 billion; total U.S. captive IC production was estimated at $3.45 billion.[2] IBM also is a technological leader. An examination of the U.S. patents issued in the area of digital logic circuits between 1969 and 1982 shows that IBM was issued 199 patents, which was more than twice the number issued to any other firm during that period. In the area of semiconductor memories, IBM was issued 194 patents between 1969 and 1982, nearly three times more than any other firm.[3] Western Electric, the captive manufacturing arm of AT&T, is the second-largest captive IC producer. In 1983, the sales value of its output was estimated at

$390 million.[4] Western Electric was the first U.S. semiconductor manufacturer to produce 256K RAMs in volume. Other U.S. captive IC producers include Delco, Hewlett-Packard, Honeywell, NCR, DEC, and Burroughs. All of these producers are large, well-established, integrated producers with a broad line of downstream products.

The field of U.S. merchant semiconductor manufacturers is more diverse though output is also highly concentrated among the larger of these firms. The largest twenty firms produce 80 percent of the total merchant output. Texas Instruments and Motorola are the two largest merchant IC producers. In 1983, these two firms accounted for roughly 32 percent of U.S. merchant shipments. Their combined sales, however, came to only slightly more than IBM's estimated 1983 IC production of $2.4 billion. The three largest merchant IC suppliers combined supplied 42 percent of the merchant market in 1983; the four largest, 51.5 percent; and the five largest, 57.7 percent. Both TI and Motorola produce ICs, based on both MOS and bipolar technology. TI leads the merchant IC industry in bipolar sales, while Intel leads in MOS sales. MOS sales accounted for 45 percent and 42 percent of total sales for TI and Motorola, respectively.[5]

In contrast with the captive IC industry, entry to the merchant IC industry has been fairly easy to the present time, and many small firms have been able to secure profitable market niches. The rate of entry increased with an inflow of new companies that specialize in the design or production of custom integrated circuits. This wave of entry was encouraged by both the increased availability of venture capital following the 1978 revision to the capital gains law and by the increased demand for custom circuitry generated by the move to VLSI technology. In addition, the availability of federal contracts encouraged the entrance of small firms to the market because relatively little capital was required to enter the market. Between 1976 and 1982, the number of firms in semiconductor and related devices industry (SIC 3674) increased 41 percent from 1,002 to 1,411; small businesses (under 500 employees) as a percentage of this total remained stable at 98 percent, according to the Small Business Administration Data Base. In 1976, small businesses accounted for 20.5 percent of total sales and 16.0 percent of total employment in the industry; by 1982, these figures had dropped to 16.2 percent and 14.8 percent, respectively. In terms of productivity, small firms are outproducing large firms; in 1982 average firm sales per employee in firms with under 500 employees was $59,000 versus $42,000 in firms with over 500 employees.[6]

Though small firms did contribute to innovations in the industry, Bell Laboratories was probably the leader in this area.

The major contribution of small firms in the development of the semiconductor industry was to diffuse semiconductor technology and stimulate competition. Diffusion of semiconductor technology occurred because the small firms with independent sources of advanced semiconductor technology exploited new markets and were able to successfully move technology from the laboratory to full-scale production.

The U.S. semiconductor industry supplies products to a wide range of end-use markets. U.S.-based firms dominate all worldwide end-use markets except the consumer IC market, which is dominated by Japan. In 1983, U.S.-based producers supplied 86 percent of the worldwide computer/data communication market, 71 percent of the worldwide telecommunications market, 74 percent of the worldwide industrial market, 36 percent of the worldwide consumer market, and 95 percent of the government/military market. In 1982, 72 percent of all ICs produced by U.S.-based firms were used for industrial applications (40 percent computer, 20 percent communications, 5 percent office automation equipment, and 6 percent other); consumer and military applications consumed 11 and 17 percent of U.S.-based production, respectively.[7] This shows a shift away from military applications, which represented almost the total demand for U.S. ICs in 1962, to computer and communications applications.

U.S.-based output also has undergone a shift away from discrete devices and an increase in the production of memories and microprocessors. In 1975, 41 percent of the semiconductors manufactured were discrete devices; by 1980, this figure had fallen to 19 percent. By contrast, memories constituted only 16 percent of production in 1978; by 1983, that figure had increased to 30 percent.[8]

The semiconductor manufacturing equipment industry is very closely tied to the semiconductor industry. The U.S. semiconductor industry, particularly the small, independent merchant firms, has drawn considerable strength from the concentration of U.S. equipment firms. There are nearly 285 American firms manufacturing semiconductor equipment, many of which have less than $5 million in annual sales.[9] Of the ten largest (in terms of sales) semiconductor equipment manufacturers in the world, eight are U.S. firms. The industry includes only a few large, established firms. The larger companies tend to concentrate on lithography equipment and automated circuit testers, which are relatively high priced products. Many of the companies in this industry have entrepreneurial roots such as Varian, Kulicke and Soffa, Materials Research Corporation, and Silicon Valley Group. In this respect, this group resembles its clients, the merchant IC producers. Though the United States has long been the leader in this industry, U.S. domination is now coming under pressure from Japan.

The worldwide semiconductor equipment market is estimated at $5.8 billion this year. As late as 1979, U.S. equipment makers controlled virtually the entire Japanese semiconductor equipment market. Now, they account for less than 40 percent of the $2.7 billion market. This is a key issue: if Japanese equipment manufacturers become dominant, this may give Japanese chip producers an edge over U.S. producers. Moreover, success in the Japanese market is vital to U.S. semiconductor equipment manufacturers. Japanese semiconductor manufacturers will spend proportionately more than their U.S. counterparts on capital equipment this year (30 percent of sales versus 25 percent). If the Japanese continue to increase their market shares worldwide, Japanese equipment purchases will soon surpass U.S. purchases.[10]

The U.S. dominance in semiconductor sales is, indeed, slipping. Japanese manufacturers are gaining market share in the United States and the rest of the world. In the world market, the U.S. currently has a 53 percent share, Japan a 38 percent share, and the remaining 9 percent is supplied by Europe, Korea, South America, and Mexico. By 1987, world market share is estimated at 49 percent for the United States, 43 percent for Japan, and 8 percent for the other group. In the United States, U.S. producers currently supply 83 percent of the market, while Japan supplies 14 percent. By 1987, it is estimated that U.S. producers will supply 77 percent of the U.S. market, Japan 20 percent of the market.[11]

Japanese firms have rapidly expanded their share of the worldwide semiconductor market in recent years. This was accomplished by focusing initial developmental and market efforts on high volume, relatively easy to produce memory chips. As of 1984, Japanese companies had captured approximately 54 percent of the worldwide 64K RAM memory chip market and 90 percent of the 256K RAM market. Recently, several of these same firms also have been making significant inroads in the markets for more complex chips, including EPROMS, semicustomized circuits, and advanced logic devices. Japanese semiconductor producers displayed impressive R&D results, particularly in the areas of very fast gallium arsenide and silicon memory devices.

Japanese firms have now surpassed their U.S. counterparts in new investment for plant and equipment, and they continue to expand capital spending very aggressively. Trade friction with the United States has been a recurrent problem as the bilateral integrated circuit trade balance has moved from a U.S. surplus in

1980 to a deficit of nearly $1 billion in 1984. Last year Japan semiconductor producers expanded their overall share of the U.S. market to about 18 percent, up from 13 percent in 1983, while the U.S. firm share of the Japanese market has remained at about 11 percent for a number of years.

The major Japanese semiconductor producers include NEC, Hitachi, Toshiba, Fujitsu, Matsushita, and Mitsubishi. These firms also are major diversified producers of computers, telecommunications systems, and other products that utilize semiconductors. Thus, the dominant consumers of semiconductors in Japan are also the dominant producers, forming a relatively closed system that is difficult for non-Japanese firms to penetrate. The vertically integrated structure of the Japanese market itself, more than any overt government policy, is now the principal barrier to greater U.S. semiconductor exports to Japan.

Direct GOJ support for the Japanese semiconductor industry is confined largely to funding for advanced R&D. Much of this R&D is carried out under MITI-sponsored computer projects, such as the High-Speed Science and Technology Computer project, where initial R&D is focused on high-speed logic and memory devices (e.g., HEMT, Josephson junction, and GaAs FET devices). The VLSI project, also known as phase 1 of the Fourth Generation Computer project, is widely credited with significantly enhancing Japanese competitiveness in 64K and 256K RAMs.

Three MITI-sponsored, specifically semiconductor-oriented, joint R&D projects are currently under way. The "Lattice Structural Elements" project aims to develop super lattice circuits with highly controlled structure and composition. The "3D Element" project aims to further increase IC integration through the miniaturization and multiplication of chip layers. Finally, the "Environment Resistant Element" project is aimed at developing circuits capable of operating under adverse environmental conditions.

Besides the MITI projects, several firms also benefit from R&D carried out jointly with NTT and, in some cases, direct semiconductor technology transfer from NTT. The semiconductor R&D carried out in NTT's laboratories is highly advanced and is generally considered to be at least on par with that carried out in the producers' own laboratories.

Formerly, the Japanese semiconductor industry benefitted from various forms of government protection, but as the domestic industry reached an internationally competitive level in the early 1970s, barriers such as foreign investment and import restrictions were gradually phased out. In fact, recent MITI efforts have focused on import promotion and encouraging voluntary export limits to reduce trade friction over the bilateral U.S.-Japan semiconductor trade imbalance.

European semiconductor producers currently are not providing a serious competitive threat to U.S. manufacturers. European chip producers' share of the world chip market dropped from 13.9 percent in 1979 to 8.5 percent last year. Over the same period, Europe's use of chips dropped from 27 percent of global consumption to 18 percent.[12] These trends do, however, show some signs of changing. The Europeans are making an effort to win back some of the home market. In contrast to the United States and Japan, which depend on their domestic computer industry to consume huge volumes of semiconductors, the Europeans have had to look elsewhere for a steady source of domestic demand.

Europe does have some world-class telecommunications concerns that it hopes will generate a strong domestic demand. Telecommunications now accounts for close to one-quarter of all chips consumed in Europe. This figure is expected to increase as countries continue to remodel their phone systems; this increase could be compounded if the Common Market is able to adopt a uniform equipment standard for all member countries. These trends could have a potentially negative effect on the position of U.S. producers, which accounted for 55 percent of Europe's $4.7 billion semiconductor market last year.[13]

French integrated circuit producers have not been strong in their own domestic market. In 1977, EFCIS, a subsidiary of Thomson-CSF, and the only major wholly French-owned IC producer, controlled only 17 percent of the French IC market. By 1981, Matra-Harris and Eurotechnique (both majority French-owned joint ventures) along with EFCIS controlled only 30 percent of the French IC market despite considerable investment in the industry by the French government during that period.[14] In the mid-1960s, the general inadequacies of the domestic semiconductor industry and the recognition of the strategic importance of this sector to the overall health of French high-technology industries led the French government to begin specific targeting of its domestic semiconductor industry.

As part of its Components Plan, the French government gave Thomson-CSF's Sescosem nearly $18 million over a five-year period to develop and produce ICs primarily for computer applications. Sescosem's technological achievements, however, were unimpressive in this area despite the public support. By the mid-1970s, France still lagged behind the United States, and increasingly behind Japan, in semiconductor technology. In 1978, the GOF launched the Integrated Circuits Plan, a five-year program specifically targeted at promoting the capabilities of IC industry especially in the area of telecom applications. Grants totalling $18 million each were given to both Thomson-CSF and Radiotechnique Compelec, Philips' French subsidiary, for R&D on integrated circuits, and $37 million to Eurotechnique for production of NMOS and CMOS, over a five-year period.[15] A number of cooperative joint ventures between French firms and U.S. firms were formed during this period to take advantage of state subsidized R&D and preferential government procurement.

To take advantage of French government support programs Matra (France) and Harris (U.S.) formed a joint venture, Matra-Harris, for the manufacture of CMOS integrated circuits and some bipolar circuits. Similarly, Saint-Gobain (France) and NSC (U.S.) formed a joint venture, Eurotechnique, to manufacture MOS circuits. Both of these joint ventures are majority owned by the French partner. Thomson-CSF also negotiated a technology assistance agreement with Motorola (U.S.) for MOS production technology.

In 1981, the French government announced the Second Integrated Circuits Plan which was similar to the first but emphasized VLSI product and process technology. This plan made $500 million available for microelectronics R&D between 1982 and 1986. In 1982 alone, $70 million was provided to the electronics industry under this program. In addition, the plan allocated $39 million to the Centre National d'Etudes des Telecommunications for the expansion of R&D directed at integrated telecommunication circuits. The plan also supports R&D into new technologies such as gallium arsenide and magnetic bubble memory.

French semiconductor equipment manufacturers also receive assistance under the Second Integrated Circuits Plan. Matra-Harris is working on a new direct wafer stepper and reactive ion etcher. The machines, developed with government assistance provided under the plan, would be manufactured respectively by Euromask (a Matra subsidiary) and CIT-Alcatel. Cameca, the semiconductor equipment manufacturing subsidiary of Thomson-CSF, is also receiving financial assistance through the plan.

In addition to providing funds for research and development, the French government has protected its domestic semiconductor industry through preferential purchase arrangements and substantially restructured the industry through consolidation and nationalization. The French military and telecommunications industries are major markets for the domestic IC producers. It is estimated that public procurement accounts for 40 percent of the purchase of semiconductors in France. Particular preference is shown to the nationalized semiconductor firms. Beginning mid-1981, the French government began to nationalize the country's major privately owned semiconductor firms. These included Compagnie Generale d'Electricite, Matra, Saint-Gobain, and Thomson-Brandt.

In the United Kingdom, most firms in the semiconductor industry have chosen to concentrate on specialized market niches, particularly in specific process technologies such as high-density CMOS and specific product technologies, particularly gate arrays and custom VLSI. These firms include General Electric Company (GEC), Plessey and Ferranti. This specialization originally stemmed from the need to produce efficiently in a relatively small domestic market. Semicustom design chips were particularly good in this respect. By the late 1970s, when semicustom design and full custom design began to gain wide popularity among end-users, U.K. producers were particularly well placed to exploit this market.

In contrast to these specialized producers, Inmos and Standard Telephone and Cables (STC) offer standard VLSI products. Inmos International was founded in 1978 with equity participation from the British Technology Group (BTG), a venture capital group within the British government. The initial government investment was $93 million in return for which the government received an equity stake. In 1980 and 1983, the government committed further funds, $60 million and $23 million respectively, to Inmos. This raised the government's ownership to 76 percent. Inmos first reported a profit in the fourth quarter of 1983; future prospects depend largely on the success of the "transputer," a computer-on-a-clip that entered volume production this year. Inmos is currently the leader in the high-performance 16K memory market. After considering alternative offers from several U.S. firms, the government recently sold its equity in Inmos to Thorn-EMI (U.K.) for $124.1 million.

The U.K. government has played a very active role in the development of its domestic semiconductor industry by encouraging R&D and promoting industry restructuring. The underlying rationale is to help foster an independent indigenous computer capability. These efforts began in the 1960s with R&D support provided to universities and government laboratories which encouraged wide dissemination of research findings. In addition, the government promoted mergers among industry firms in the late 1960s, offering funds to newly merged firms. After 1967, the government also provided funds for production R&D.

More current initiatives include the five-year Microelectronics Industry Support Program (MISP) begun in 1978 and the three-year Microelectronics Applications Project (MAP) also begun in 1978. The MISP program, also known as the Alvey Advanced R&D Program, provided $200 million to subsidize collaboration research principally by GEC, Plessey and Ferranti into VLSI fabrication technology. In 1984, the MISP program was extended to 1990. The MAP project was targeted at promoting industrial use of microelectronics and the development of new microelectronics-based products.

West Germany is by far the largest producer of semi-conductors in Western Europe. Much of its production is, however, generated by foreign-owned firms. Indigenous semiconductor producers include Siemens, AEG-Telefunken, Robert Bosch and Grundig. The first two supply to the open market, and the latter two produce almost entirely for internal consumption. All the West German producers are highly diversified, vertically integrated firms. Siemens is clearly the dominant producer. Its production is divided almost evenly between ICs and discrete optoelectronic devices; its IC production consists of 40 percent MOS and 60 percent bipolar. Because of its strength in telecommunications equipment, Siemens is also a major producer of telecommunications chips. In addition, the company is strong in power MOS semiconductor products.

Over the last ten years Siemens has undertaken large investment (roughly $200 million) aimed at closing the gap in VLSI technology with its foreign competitors. A sizeable amount was provided by the German government. As part of this endeavor, Siemens built two VLSI centers. In addition, Siemens has become involved in the Japanese and U.S. markets through investment in local semiconductor firms and joint-venture arrangements.

The West German government has actively targeted its domestic semiconductor industry since the mid-1970s with the establishment of the four-year Electronics Components Program which provided funds to universities, research firms, and industry for research and development projects. The program was continued and enlarged as the Microelectronics Program. It aimed at strengthening the country's position in semiconductor production by supporting R&D. In addition, in 1982 the government announced it would fund a three-year development program for marketable microelectronics-based products at DM450 million. The Federal Ministry for Research and Technology (BMFT) also targeted funds to specific companies, such as Siemens and AEG-Telefunken, for research on electronic components.

More recently, in acknowledgment of the two-year lead in gallium arsenide (GaAs) technology held by the United States and Japan, the West German BMFT has announced a four-year, $40 million research and development program to begin in early 1986. The aim is to help the country's semiconductor producers capture a 10 percent share of the world's GaAs market by the end of this decade. The program is designed to coordinate the activities at semiconductor companies and universities under the direction of the Fraunhofer Institute for Applied Solid State Research.[16]

As a result of extensive targeting in the semiconductor industry, competitive conditions have been significantly altered in some countries. This was illustrated recently in the Japanese semiconductor market by a photoaligner product known as a stepper. An advanced stepper was introduced by an American company in 1977. Since steppers can significantly raise the yield of complex chips on a wafer, the demand for this product has been very strong. In 1979, American manufacturers held an 80 percent share of the world market for photoaligners. There was no significant Japanese competition in the market until 1981, when Nikon introduced its first stepper to the market. Nikon's product was developed as a result of MITI recommendations. MITI and five Japanese semiconductor manufacturers, NEC, Hitachi, Fujitsu, Toshiba, and Mitsubishi Electric, contributed a total of $285 million for development contracts directed at the equipment industry. In addition, MITI encouraged Japanese semiconductor manufacturers to purchase Nikon's equipment. By 1984, Nikon held about 50 percent of the domestic market.[17] In addition to photoaligners, MITI has heavily supported other equipment makers, especially in the area of automatic testing equipment used both in research and development and production. Given MITI's aggressive targeting policies in this sector, it is not surprising that U.S. equipment makers' share of the Japanese market has dropped from 70 percent in the late 1970s to 30 percent today.

A wide range of nontarrif barriers has been encountered by U.S. semiconductor and semiconductor equipment manufacturers in Japan and Western Europe. The nontarrif barriers most commonly encountered by the companies that were interviewed included: standards, preferential government procurement, difficulty in obtaining patent protection and financing for establishment of overseas operations, and government imposed performance requirements such as encouraging U.S. firms to transfer technology to local companies.

Problems with standards were encountered in Japan and West Germany; examples of these difficulties can be found in Case Studies No. 6 and 7. In Case No. 6, the company experienced difficulties selling to NTT, where specifications are used to discriminate against non-Japanese products. This same company encountered similar problems selling to the European PTTs; specifications were altered to favor local products.

Difficulty in obtaining patent protection in Japan is shown in Case No. 7. Once the company's Japanese competitors began to reach a certain technological level, the company began to experience delays in granting of protection and, in some cases, an inability to obtain protection. Two of the companies interviewed

(Cases No. 7 and 9) encountered difficulties obtaining financing in Japan during the 1970s. However, more recently a company (Case No. 10) was able to obtain financing from the Japanese Development Bank, which indicates that this may no longer be an important market barrier.

Table 6.1 outlines the various strategies undertaken in the face of these market barriers and summarizes the key reasons given by the interviewees for the strategies that were chosen.

TABLE 6.1
Summary of semiconductor case studies

Firm No.	Total Assets ($ Mil)	Strategy	Reason
6	over 100	Joint venture in Austria for manufacturing and design; Design Center in Japan	Circumvent duty system; enable interactive custom design; difficulties selling to European PTTs (specifications used to favor local vendors); broadly defined NTT specifications; nationalistic attitudes
7	96	Joint venture in Japan since the 1950s for sales and marketing only (not very successful); hope for wholly owned manufacturing in Asia in the future; no great optimism for European market	Unable to obtain patents in Japan; problems with "slippery" standards in Japan and Europe; nationalistic attitudes in Europe and preference for Siemens in Germany
8	265	Wholly owned sales and manufacturing subsidiaries employing local nationals	Strong commitment to "making it on its own" and desire to maintain control
9	25	Joint venture in Japan for sales, manufacturing, R&D for local market; wholly owned subsidiaries in Europe	Problems with distributor relationships in Europe (copied product) and Japan (no contact with customer); difficulty obtaining financing for 100 percent owned facility in Japan; received pressure from French government to form a joint venture with local firm and transfer technology
10	65	Began with trading company in Japan and later established joint venture for sales and manufacturing; wholly owned sales subsidiaries in Europe with manufacturing site in France	Need to have more direct involvement in Japan because of market size and importance; difficulty hiring salesmen and engineers in Japan; unable to participate in MITI-funded research

(Continued)

TABLE 6.1 (Cont.)

Firm No.	Total Assets ($ Mil)	Strategy	Reason
11	6	Licensing agreement in Europe that resulted in a breach of contract; licensing agreement in Japan that had to be terminated; currently using distributor in Japan	Loss of technology through licensing agreements in Europe and Japan
12	300	Originally used trading company in Japan for distribution and service; in 1980 established joint venture in Japan for service, with intent to develop sales and manufacturing there; in Europe sold through distributors until 1973, then established a sales and later manufacturing subsidiary in Europe; in 1981, established joint venture in France for R&D manufacturing and sales	MITI-funded development of competing technology given to Japanese firm and Japanese customers urged to buy from this local company; Japanese targeting of domestic industry; Japanese customers began to require customization of product that could not be provided from U.S. manufacturing site
13	190	Joint venture in Japan that resulted in loss of technology; after this experience, used a local distributor in Japan until recently, then a joint venture for sales and manufacturing was established	Loss of technology through a joint venture in Japan; as Japanese catch up technologically, company must provide more and more modification of product; inability to hire work force unless affiliated with large company in Japan

CASE NO. 6

The firm is involved in the electronic components and materials business. It manufactures custom, semicustom, and standard MOS/VLSI (metal-oxide-silicon, very large scale integrated) circuits. The company is headquartered in California, with auxiliary sites in the Northwest, Austria, and the Philippines.

The company has been in the custom MOS/VLSI business for over eighteen years and, as of fiscal year 1984, has produced more than 3,000 different specialized circuits for computer manufacturers, telecommunications companies, automobile manufacturers, and consumer product companies. According to its most recent promotional materials, it offers a continuum of customized

MOS/VLSI solutions, from semicustom gate array circuits and standard cell designs, to interactive full custom circuits, all of which are tailored specifically to the customer's product. Thus, it is necessary that the firm work very closely with its customers.

The company's custom chips offer a wide range of advantages over standard components: total system costs can be lowered as a result of reductions in the physical size, power requirements, and assembly steps required; product reliability and production yields can be increased through the elimination of potential failure sources; performance advantages and specialized functions unattainable through discrete logic are offered. Custom chips do, however, require a reasonable amount of lead time. If lead time is unavailable, the firm offers semicustom gate arrays that will get the customer into production fast with low development costs. It offers the customer a wide range of flexibility in the design of its ICs; the customer can be involved as much or as little as desired.

Since 1984, the company has had a design center in Japan. The design center is managed as a joint venture between the company and a Japanese partner. The Japanese partner is in the petrochemical business; although it does have some interests in other areas of electronics, it is not a competitor in the field of semiconductors. The arrangements included the U.S. firm transferring its design technology and the Japanese partner sharing the capital investment for the venture. With the U.S. company, the joint venture has access to all design technology. The products currently being designed at that facility include gate array circuits, full custom circuits, and standardized products for the Japanese and Asian markets. The Japanese partner was given rights to sell primarily to the Asian market. The U.S. firm has experienced little difficulty obtaining patents in Japan.

At the present time, all manufacturing generated by the design center is taking place in the United States and being shipped back to Japan and distributed through the Japanese joint venture. In the long run, consideration is being given to the potential need for in-country manufacturing facilities. Given the company's emphasis on customized products and services, interactive design is key. As the volume of work at the joint venture increases, the need for interactive design will become more urgent.

One of the primary markets for these products in Japan is telecommunications. The company has experienced some difficulty with NTT with regard to product standards. The specifications in the NTT specification book are defined so broadly that the company has had to send design personnel directly to the NTT central office in Tokyo to ask the particular inspector involved for his interpretation of a specification before submitting a product for approval. Unfortunately, the company has found that even these efforts, which have cost additional time and money, do not ensure that the specification will be interpreted in the same manner when the product comes into the NTT for inspection. The company perceives that these broadly written specifications are used to discriminate against non-Japanese products. Although it does not believe that these difficulties reflect a general NTT policy, the NTT inspection system and broadly written specification books seem to allow for strong nationalistic attitudes on the inspection level to work against non-Japanese companies.

The company's approach to the European market has been somewhat similar. Since January 1982, the company has had a joint venture in Austria. The joint venture is a full-fledged manufacturing facility capable of producing the firm's latest state-of-the-art technology. This has been helpful to European marketing efforts, in general. However, the company has found that the joint venture in Austria has afforded them no advantages in selling to France or Germany, for example, other than lowering costs as a result of reduced freight charges and custom duties.

In Europe, the company sells directly to the local postal authorities. Although standards are set in Europe by CCITT, it has found that the specifications are subject to interpretation by a regulatory agency that may favor nationalized manufacturers. Specifications often appear to be interpreted on a case-by-case basis, often to the disadvantage of nonlocal manufacturers, and new product specifications appear to arise as a result of research and development being performed by local companies.

CASE NO. 7

The firm is involved in the design, manufacture, and sale of power semiconductor products. The company is a major supplier of high-power discrete semiconductors and solid-state relays with industrial and military applications. In the mid- to late 1970s, it began intensive research and development efforts to apply MOS process technology to power semiconductors. The result of its research and development program was the introduction in 1979 of a line of power devices manufactured by use of integrated circuit MOS technology.

The company's power device was introduced in the late 1970s as an alternative to the bipolar transistor. Both devices have applications in various types of equipment, including computers and computer peripherals, telecommunication equipment, audio amplifiers, ignition systems, switching power supplies, radio transmitters, and motor speed controls. The company introduced its power device in 1979 by employing state-of-the-art integrated circuit MOS techniques. The devices are now used in a variety of higher-power applications previously reserved to the bipolar transistor. Because this power device cannot be used in place of the bipolar transistor without changes in a product's circuitry, the market could expand only as rapidly as new products were designed or appropriate changes in existing products were made. Despite this limitation and reductions in price, its product sales have increased substantially over the past three fiscal years from roughly $12 million in 1982 to $33 million in 1984.

Success in the international market is critical for the company. The company's revenues for 1984 were more than $140 million worldwide; approximately 30 percent of those revenues were generated outside the United States. Currently, the principal foreign market for these power devices is Europe. This past year, roughly $50 million in revenues was generated in the European market. Although approximately 50 percent of what it sells in Europe comes from the United States, the company assembles and manufactures in Europe, and has a distributor group in Europe that also distributes the products of several other companies. A large fraction of its customer base in Europe is maintained on a noncompetitive basis because of military applications. Almost all the major companies are customers including: Ericsson, Mueller, Gould, M. D. Messerschmidt, BBC, and IBM. Its only really effective competitor in Europe is Siemens.

Within the overall scheme of its goal to penetrate the international market with its power device, Europe is not expected to play a major role. The company has found it very difficult to sell into these markets. This problem is particularly acute in Germany, where Siemens manufactures a competitive product. Siemens' products compete fiercely with the firm's comparable products in Germany even though they pose no serious threat to the firm outside of Germany. The company has encountered preference for Siemens' products at Bosch, for example. In this case, it is hard to differentiate between nationalistic attitudes and simply a preference for a local manufacturer who can guarantee an uninterrupted supply of goods and a strong local service organization.

It also has experienced problems selling one of its circuits directly to Siemens. Siemens required a very large number of the circuits; the order would have resulted in a million-dollar annual business. After Siemens had bought some of its samples, the specification for the product was abruptly changed and the business disappeared. Shortly after, Siemens began manufacturing the circuit itself. Experiences such as this have left the firm very skeptical about its ability to make any further inroads to the market and have discouraged the company from viewing Europe as a strategic growth market in the future.

The firm views Asia as essential to its future growth and Japan as the key to that market. Competition in the power device market in Japan is well under way. Japan's power device suppliers began a vast stepping-up of production last year in response to increased customer interest. Its goal is to supply these devices to the United States in large quantities. Major Japanese suppliers of these devices include NEC, Hitachi, Toshiba, and Matsushita Electronics; they still do not have a large production base for their devices in Japan. The Japanese hope to use their expertise in fine-pattern, VLSI technology and memories made with the same MOS technology to make them a major force in the power device marketplace.

There is general agreement that the Japanese market is beginning to incorporate these power devices into product design on a large-scale basis. Toshiba is expected to be the most aggressive in taking advantage of the local market's new interest in the high-frequency, high-efficiency device. Toshiba expects to double production each year to account for 10-20 percent of its power device production, according to a company source. All the other Japanese producers are following suit. At present, Japanese production of this device is equivalent to 1 to 2 percent of bipolar production. In the next five years, sources indicate that Japan's power device production will be divided evenly between the two. Unfortunately, despite the excitement over potential applications of the power device in Japan, the company, which introduced the device to the Japanese market in 1979, has been unable to increase its market share beyond 10 percent.

The company has been involved in the Japanese market since the mid-1950s, when it began a joint venture there. At that time, it owned a substantial percentage of the venture. It now owns only a small percentage. The joint venture sells the firm's products as well as products of its own manufacture in Japan, Korea, Taiwan, and other countries in Southeast Asia. It is a publicly held company traded on the Tokyo Stock Exchange; the largest stockholder is a Japanese corporation that manufactures electronic equipment.

In terms of return on investment, the Japanese joint venture has been very successful; however, in terms of an entree into the Japanese market, the joint venture has not fared as well.

In 1974, in an effort to deburden itself from the financial problems associated with running a company that was half U.S.-owned and half Japanese-owned, which could not take advantage of the tremendous debt-equity ratios that are available in Japan, the company decided to put the joint venture on the New York Stock Exchange and the Tokyo Stock Exchange. However, it encountered problems listing the joint venture on the Tokyo Stock Exchange. After a considerable delay in listing, the company made some inquiries. It was informed at that time that prior to a listing on the Tokyo Stock Exchange, the head of the exchange had to visit the plant in a formal ceremony. The company then sent a formal invitation to the head of the Tokyo Stock Exchange. After several weeks, there was still no response. It then made additional inquiries and was told that Mr. X of the Tokyo Stock Exchange felt it would be a loss of face if he were to attend a formal ceremony at a U.S.-managed company. In fact, the joint venture was far from a U.S.-managed company; there was only one U.S. manager present on the board of directors. Nevertheless, this confrontation came down to a "stand-off," and the company was forced to remove its one U.S. manager. Not surprisingly, the joint venture was listed on the Tokyo Stock Exchange within the same month.

The company also encountered problems with specifications. One particularly frustrating experience took place with a sale to Toshiba. In the beginning, Toshiba presented a specification to the company. It was able to meet that specification and a very large quantity of parts were ordered by Toshiba. After the order was placed, Toshiba began to stall the sale. By regular mail, the firm would receive lists of problems found with the product and requests to investigate the problem and report back to Toshiba with its findings. This went on for several months, during which company engineers went back to Toshiba with evidence that approximately 80 percent of the problems found by Toshiba with the company's product has been Toshiba-induced errors or mistakes. Shortly after that, Toshiba cancelled the entire order and manufactured the device itself.

Futhermore, the company has been unable to obtain patents for its technology in Japan, although it now has over fifty patents in countries around the world. It has filed for a number of patents in Japan and has waited as long as four years for the first action. The Japanese examiners have been claiming that there is prior art in Japan. Nonetheless, the company was able to obtain patents for the same technology in the United States and Germany and was able to prove that there was no prior art. The company appealed these findings of prior art with the Japanese patent office and was overruled. It still has no patent protection for its technology in Japan. Fortunately, the technology is well patented in the United States; this will matter greatly when the Japanese power devices begin to hit the U.S. market.

One example in particular, however, demonstrates on a broader scale the dynamics of the difficulties encountered in Japan. In 1981, a group of its engineers visited the NTT design laboratory on a routine matter. At that time, NTT was busy designing what would be a high-volume telecommunications product. The company's engineers observed that the new design was incorporating two Hitachi transistors rather than using only one of the company's power transistors. At that time, there was a frank admission by NTT employees that this clumsy design was being used so the product could utilize Hitachi technology. Through their own admission, it was a grossly inefficient design. The company had a product that was twice as efficient as the transistors produced by Hitachi; nonetheless, NTT would not incorporate its transistor in its design. In doing this, NTT demonstrated its determination to wait until there was a Japanese source before using the power device in its designs.

The company's power device has been on the Japanese market since 1979. Even though it has been able to capture the leading position in this technology in the United States, the market for this power device has not developed significantly in Japan. The market for the device in Japan is one-hundredth the size of the U.S. market for the same device; given Japan's technological strength in products that could utilize this new power device, this does not make sense.

Although none of the difficulties experienced in Japan could be directly related to a specific government policy toward the domestic semiconductor industry, these problems are nonetheless typical of the sort of difficulty experienced by other companies trying to sell semiconductors in Japan. It is clear that the company's difficulties in the Japanese market have resulted in a very low market penetration. Moreover, the company's market share has remained stable at 10 percent for nearly six years despite its efforts to the contrary and its obvious leadership position. The observed stability of market shares over time leads one to believe that there is a certain amount of parcelling-out of the semiconductor market in Japan, especially in areas of head-on competition between firms.

Despite impediments encountered in Japan, the company plans to pursue that market with greater intensity. Although the current market for this power device in Japan is approximately $10 million, the market is expected to skyrocket during the next few years. The strategy it intends to pursue for a renewed attack on the Japanese and Asian market is to build in Asia what is sold in Asia.

CASE NO. 8

The firm is a twenty-year-old publicly traded company specializing in the manufacture of linear integrated circuits (analog ICs). In the field of data acquisition, the company is one of the largest suppliers, with relative competitive strength built on technical innovation and support, quality and reliability, worldwide manufacturing strength, and a broad range of products. The company's overall strategy to consistently focus on the development, manufacture, and marketing of precision ICs used to process real-world signals has proved highly successful; real-world connections play an increasingly important role in the expanding market for microprocessor-based high-technology tools to boost productivity and competitiveness. In 1984, the company generated approximately $300 million in sales worldwide. Today, roughly 40-45 percent of the company's sales are generated overseas.

The company first became involved in international markets in 1966 when it opened direct, wholly owned operations in the United Kingdom. At this time, the company was only about a $1-million-a-year business. It then followed this same formula in West Germany in 1968, France in 1969, and Japan in 1970. Every year or so after that, the company has added another direct, wholly owned operation overseas. The company now has nine of these operations in Europe, which include: Belgium, Denmark, England, France, West Germany, Holland, Switzerland, Italy, and Sweden. With the recent addition of Israel, the company has eleven overseas sales subsidiaries in total. The company has set up each sales subsidiary by copying the pattern of the local distributor; it builds organizations of local nationals who distribute to their own markets. As a matter of corporate policy, the company did not believe in the viability of the joint-venture arrangement. The company had a strong commitment to "making it on its own" and it did not want to lose control of its overseas markets to a greedy partner. Company executives adopted a long-term approach to the market. If a country showed promise of evolving into a significant market for the company's products, the company targeted that country for a wholly owned subsidiary. The company would then wait, sometimes for several years, until it found the right individual to run the organization. The company believes that with the right leader, it is easy to attract other talented people to the firm.

The European market is, by far, the company's largest overseas market. Roughly 30 percent of the company's current sales are generated in the European marketplace. The company's list of European customers is impressive and includes Siemens, Philips, Marconi, Olivetti, ITT Belgium, Aerospatiale, Thomson-CSF, and Smith Industries. These European sales subsidiaries are all self-supporting and doing respectably well within their own marketplaces. The subsidiaries function as local distributors, buying the product from the company and selling it at arms length. The company has encountered no discrimination in Europe. It attributes this to the fact that it has a unique product line that does not have strong direct local competitors.

In 1971, the company decided to establish a manufacturing facility in England. This facility manufactures a hybrid product line including mechanical linear and rotation converters. These products are designed, developed, and manufactured for worldwide distribution. All research and development and engineering functions are accomplished at the manufacturing site. In 1975, an integrated circuit facility was established in Ireland. This design site manufactures and markets integrated circuits also for worldwide distribution. There were many financial incentives for building a manufacturing site in Ireland, including tax holidays, training aids, and various capital investment incentives. Both England and Ireland have good skilled labor forces and, as members of the Common Market, are not subject to the 17 percent duty on integrated circuits. The company now has eight manufacturing facilities around the globe, each producing its own distinct product line that it distributes worldwide. In the long run, the company hopes to evolve these sites into duplicate manufacturing sites producing the same full-product line. This would provide back-up supply and allow the company to derive maximum benefits from tax incentives and duty-free trading zones.

In line with its general approach to the overseas markets, the company never seriously considered a joint venture in Japan. The company established a wholly owned subsidiary in Japan in 1970, which distributed the company's products to Japan and the Far East. By its first anniversary, the Japanese subsidiary had already generated $1 million in sales. Today, the company generates roughly 11 percent of its worldwide sales in Japan. The company's customers include Hitachi, Toshiba, Mitsubishi, Matsushita, and Fujitsu. The biggest problem encountered in Japan was attracting qualified personnel to work at the company. The company had to work very hard to establish a solid reputation in Japan. Each year, the company hired a handful of individuals and trained them from scratch. Little by little, the company built up its work force.

The company always feared that if its business grew too large in Japan, MITI would try to "reclaim" that business for Japanese firms. However, as the company continued to grow, no such problems emerged. The company continues to have very little local competition in Japan. The company attributes this to the fact that the product line is very sophisticated, the manufacturing process is very difficult, and it is not a large-volume business.

Within the last few years the company established a manufacturing site in Japan. This facility, for the most part, does assembly work for the Japanese market only. However, some R&D and manufacturing is done there. The site was established because the company perceived that its Japanese customers felt more secure about the supply and reliability of the product if it was manufactured in-country. The company tried to copy the Japanese style of manufacturing as much as possible. The Japanese factory utilizes Japanese labor and Japanese manufacturing equipment. The company is particularly satisfied with the Japanese semiconductor manufacturing equipment used at the factory and is considering importing the equipment for its U.S. facilities.

Overall, the company has been pleased with its performance in Japan. The company is especially optimistic about recent U.S.-Japan dialogues. It has already been approached by several of its major Japanese customers with requests to find ways to increase these companies' purchase levels. These customers are going out of their way to increase their current volume of purchases from the company because of Nakasone's recent mandate to increase purchases of U.S. goods. If these firms actually follow through with purchase orders, the company's Japanese sales could increase substantially this year.

CASE NO. 9

The company has been in business since 1959. Though originally a supplier of instrumentation to the aerospace industry and, in particular, the Apollo project, the company pioneered the application of mass flow technology to the needs of the semiconductor industry and manufactures the industry's most widely used mass flow meters, mass flow controllers, gas blenders, and process controllers. It developed and currently markets its own innovative diffusion furnace system and photochemical vapor disposition reactor.

Since 1970, the company has been involved solely in the semiconductor equipment manufacturing industry. Recognizing that the semiconductor industry would define the leading-edge of technology to become the "crude oil of the eighties," the company aimed to develop new and unique tools for the industry that would revolutionize the way semiconductors are made. To this end, the company has achieved success. For the first five years, the company had no competition whatsoever. Adequate patent protection was obtained early in the United States, Japan, France, Germany, and the Netherlands. As a result, the firm was able to consolidate large market shares well before competitors had a product. Later, the company grew by branching out concentrically from mass flow meters into related technologies and processes. Often, the company invented a new process that replaced a function previously performed by a customer or supplier.

The company made its first foray into the international markets when it attended an instrumentation sales exhibit in London that was set up by the U.S. Department of Commerce. At this exhibit, the company chose a British distributor for its product. After a year of correspondence and very few results, the company found another, better equipped distributor to do the job, one more closely attuned to the product market. This Dutch company then became the distributor for the new flow controls that the company just invented. The company shipped parts to the distributor, which then assembled the parts, calibrated the instrumentation, and distributed the final product in Europe. Slowly, however, the distributor became a manufacturer and a competitor. As this became obvious to the company, the distributorship agreement was terminated and a wholly owned company was set up.

In the meantime, the Dutch company had duplicated the equipment and was selling it in Europe. The company did not believe it could win a suit against the Dutch firm in the European courts. Eventually, however, the Dutch firm tried to sell the duplicated product in the United States. At this juncture, the company went to the International Trade Commission to file suit. One week before the case was to be heard, the Dutch company decided to settle out of court; the settlement was very satisfactory to the company. Nonetheless, the suit had already cost each firm roughly $500,000 in legal fees.

In 1972-1973, the company decided to introduce its product line in Japan. The company was still very small, generating about $5 million in sales. The product was essentially unknown in Japan at that time. A small Japanese trading company was found, and it started selling the company's products in Japan. Almost from the beginning there were problems. The company could never find out what the trading company was charging for the products or to whom the trading company was selling. In fact, the trading company was charging twice the U.S. price in Japan and making a very handsome profit. The Japanese recognized the significance of the company's innovations much faster than did the Americans. Business became better and better in Japan, but the company was kept isolated from the customer and the market. It soon was apparent that if the company did not want to lose the business, it had to have a joint venture in Japan.

Since the company had no trust in the Japanese trading company with which it had been doing business, it decided to form a 100-percent-owned company. The Japanese had just announced that they would allow 100 percent foreign ownership of companies in this particular technology intensive field; this was in the mid-1970s. After spending a year trying unsuccessfully to get financing for the venture, the company decided a 100-percent-owned company would be impossible. A joint venture was formed with a young Japanese salesman, and the ownership was divided evenly between the two parties.

This joint venture has been very successful; it is the envy of the industry and touted by the Japanese government as an example of the way joint ventures should be run. The venture has always made a profit; the company receives 4 percent royalty of sales and half the profits. The joint venture is growing at 15-20 percent a year. In addition to its sales capacity, the joint venture has evolved into a full-fledged manufacturing facility with completely independent research and development capabilities. Though the R&D is directed toward the Japanese market, some of its products have served the U.S. marketplace as well. The joint venture itself, however, was limited by the original agreement to selling in Japan. Eighty percent of its sales are in flow controls, roughly $15 million a year.

For the first six or seven years, the company had no competition in Japan. Recently, however, a Japanese company was formed for the sole purpose of competing with the company. It has purportedly been backed by a consortium of large semi-conductor companies including Hitachi, NEC, Fujitsu, and Toshiba. So far, the company still has 65 percent of the market in Japan. The company does fear, however, that it may now experience some problems obtaining adequate patent protection for its new flow controllers, although it has not seen any evidence of this so far. Because the worldwide market for the flow controller is only about $50 million, the company believes that the Japanese government has no specific targeting programs directed toward this product sector. To combat competition, the company has pursued in the past and continues to pursue an aggressive policy toward innovations in its product lines. Research and development is performed on an ongoing basis. At any point in time, the research lab is already well into the process of obsoleting a product by the time it is introduced to the Japanese market.

The company has been successful in Europe; it currently has an 80 percent market share there. Wholly owned subsidiaries were established in West Germany in 1978, France in 1981, and the United Kingdom in 1983. The company encountered no difficulties establishing the subsidiary in West Germany; it has local German management and employs about twenty-five people. Since nationalism in Europe often makes sales between members of the community difficult, the company opted to establish wholly owned subsidiaries in France and the United Kingdom.

Problems were encountered in France. When the company applied to the French government for permission to establish a wholly owned subsidiary in France, it was asked to find a French company with which to form a joint venture. The company was to offer its technology and financing but allow the French partner to own 51 percent. The company was not interested in an arrangement of this sort. For the next year and a half, the company continued to work with a banker and attorney in France and eventually was able to form a 100-percent-owned subsidiary in 1981.

From the beginning, the company set out to develop a worldwide customer base. The company's goal has been to supply each market area with nationally identified products sold and serviced by local personnel. In Japan, for example, the joint venture designs, develops, manufactures and services a line of products specifically aimed to fill the needs of Japanese customers. The company believed that in order to compete effectively in the Japanese market, it would have to offer locally built and serviced products made to local standards. Where possible, the company has opted for wholly owned subsidiaries; it believes that control of overseas operations is essential. The company attributes its success in the international market to its early technological lead, which allowed it to come to market with needed products years before its competitors. As a result, the company was able to dominate the market for its products during the nascent stages of the semiconductor equipment manufacturing industry. This large market share made it difficult for other companies to successfully compete against the company.

CASE NO. 10

The company is a twenty-eight-year-old publicly held firm that provides materials and manufacturing equipment primarily to the semiconductor industry. Approximately two-thirds of the company's revenues are derived from the semiconductor industry; the other third comes from the telecommunications and computer industries and other sputtering applications. Within this broad customer base, over 99 percent of the applications of the company's products involve thin film technology. Although the company's manufacturing technologies are very different-- equipment, materials, and ceramics--the applications of its products all relate to this one technology. In 1984, the company generated nearly $100 million in sales. The company's main products are high-technology consumable materials from which thin films are made, high-performance ceramic substrates on which hybrid integrated circuits are deposited, and sputtering and plasma etching equipment for thin film fabrication.

The company began selling in Japan in 1962 through a small trading company that acted as the company's exclusive distributor in Japan. Beginning in the early 1960s, the company established five wholly owned sales subsidiaries in Europe. These were established in France, West Germany, the United Kingdom, the Netherlands, and Italy. In 1976, the company built its first full manufacturing site overseas in Toulouse, France.

By 1980, the company had witnessed the semiconductor equipment manufacturing industry go from almost total U.S. domination when it first began in the business to an industry comprised 45 percent by U.S. firms and 35 percent by Japanese firms. The company began to recognize at the same time that its largest potential market was Japan. The company was convinced that the size of the Japanese market and the MITI-coordinated thrusts into specific areas militated against technical dominance by a U.S. company that was not also an integral part of the Japanese domestic scene. Heretofore, most Japanese production equipment used for semiconductor manufacture were U.S. produced. In the early 1980s Japanese firms began to come onto the domestic (Japanese) market with more reliable products, which were competing heavily with the U.S. imports. The company decided it had to establish full operations in Japan or it would eventually go out of business as the pattern was clear-cut-- once Japanese suppliers satisfied their domestic market, they would begin exporting to the United States and Europe.

The company chose to establish a Japanese company and commence direct sales and service, and to make an immediate investment in a plant for manufacturing equipment in Japan. The company chose to set up its Japanese operations as a joint venture in which it had 80 percent ownership, and the remaining 20 percent was given to the small trading company with which it had been doing business for nearly twenty years (just purchased this minority interest August 1985). This decision was based partly on the personal relationship between the head of the company and the head of the Japanese trading company and partly on the belief that a Japanese partner, even very small, might help the company gain acceptance more readily in Japan.

The company approached MITI with its plan to set up a Japanese company to manufacture and sell semiconductor manufacturing equipment and materials. MITI offered no assistance whatsoever to the company. It was evident that the company would have to fend for itself. The company held a large press conference in Tokyo during which it described its product line and its interest in setting up a manufacturing site on the island of Kyushu (Silicon island) in Japan, and asked to be welcomed by the local prefects. The company soon received invitations from several prefects to bring its business to their area. Several meetings were set up, and the company chose to settle in an area that was headed by a prefect who used to work for MITI and was very well connected in both the political and financial channels in Japan.

The company then began to look for land on which to build its plant. It was shown several desirable locations; however, the prices were astronomical. The company then approached the local prefect and explained that the land was prohibitively expensive and asked what kind of deal could be made. The local government then bought the land from the private owner and offered to rent it to the company for $30,000 a year for three years, at the end of which it had the option to buy the land. The total cost of the land now owned by the company was around $150,000--far cheaper than the original estimates.

The company then had to find financing for this venture. Although it had never been done by a U.S. firm before, the company decided to seek financing directly from the Japanese Development Bank (JDB). After eight months of negotiations with JDB and no results, the company sent a telegram to the JDB office in Tokyo stating that company executives would arrive on the following Wednesday to finalize the loan. The company then spent the next few days skirting phone calls and telexes from surprised JDB officials and arrived the following Wednesday at JDB offices in Tokyo. Miraculously the loan was granted. A large press conference was held by the JDB and the company immediately following the signing of the documents. The loan was viewed as a great coup by the U.S. business community.

It was clear to the company that certain endogenous elements had come into play. First, the local prefect had some personal connections at the JDB and had interceded on behalf of the company. Second, the JDB, along with the Japanese government and MITI, wanted to make an example of the company to help ease nascent tensions between the United States and Japan over the closed nature of the Japanese market.

The company still had obstacles to overcome. The greatest of these was hiring engineers and salesmen. The company has no access to graduates of the top Japanese universities, but it does hire several graduates a year from lesser-known universities which it takes pains to train. Fortunately, the company was able to hire eight of the ten engineers who handled the company's accounts for the trading company under a complicated loan and acquire arrangement. The company also was disappointed by the failure of its efforts to get involved in government-sponsored research. Although the research money involved was insignificant, the company was hoping to benefit from the government's "stamp of approval." After spending time and effort unsuccessfully pursuing this avenue, the company simply gave up.

Today the company generates roughly 15 percent of its total sales in Japan, and this represents a threefold increase. It is anticipated that sales could reach 25 percent of world totals in Japan. The company believes its approach to the Japanese market has been successful. The company continues to believe that an American company cannot do business abroad effectively through a trading company. Because the company's customer base was diminishing in the United States and enlarging in Japan, and the nature of the company's products required a closer interaction with the customer, it was clear that the company had to commit more heavily to the Japanese market if it was going to be successful. One of the surprising by-products of the venture is that manufacturing efficiencies are so good in the Japanese plant that product is now being shipped from Japan to Korea, the People's Republic of China, and Europe.

CASE NO. 11

The company is a small, privately held firm that manufactures three types of wafer-slicing machines used in the manufacturing of integrated circuits. The company recently introduced the only computer-controlled saw in the world capable of cutting 8-inch-diameter silicon. In 1984, the company generated $7 million in sales. The company enjoys an 80 percent market share in the United States. Its machines cost between $80,000 and $130,000 per unit. Last year, the company generated 15 percent of sales outside the United States.

In the mid-1970s, the company established a licensing agreement with a German firm that manufactures ball bearings, grinding spindles, vertical surface grinders, and other machine tools. The German firm was licensed to manufacture the company's wafer-slicing saw. The German firm agreed to abide by U.S. law; the agreement is in effect until 1992.

In 1982, the company began to sense that the two firms were drifting apart and to suspect that the licensee was gearing up to become a competitor. A meeting was set up with the managing director of the German firm. The company stated its suspicions at that time and stressed that if the two companies did not come to an understanding, they would evolve into competitors and there would be a breach of contract. The company recommended that the licensing agreement be converted into a joint-venture arrangement in which the company would have some equity. The German firm said this was a good idea. Nevertheless, after several months of correspondence, the two firms had not come to an agreement over the establishment of a joint venture.

By this time, the German firm had designed a slicing machine of its own, which utilized the vertical ingot that was not the company's design. The German firm claimed that the new machine did not fall under the original licensing agreement and did not want to pay the company royalties. However, the new machine did utilize the company's pivot-arm technology. Since the agreement covered any machine based on the company's technology, the company is now contemplating taking the German firm to court for breach of contract. The German firm has been pressing the company to release the firm from the original agreement, and recently released its new slicing machine without the company's permission.

The company had a similarly dissatisfying experience with another licensing arrangement. In 1974, the company licensed a Japanese firm to manufacture its 16-inch slicing machine. The machine was exhibited at the U.S. trade center in Toyko that year and received a lot of publicity. The Japanese firm left the exhibit with the machine in tow. After six months the company had still not been paid in full. The company then introduced a larger 22-inch slicing machine. The Japanese licensee then contacted the company asking for the technology for the larger machine, claiming that the agreement ensured the transfer of all improvements on the original machine. The company told the Japanese firm that it wanted the money it was owed first. The company received its money within forty-eight hours. The company terminated the agreement shortly after. The Japanese firm is not a substantial competitor today.

Since the early 1980s, the company has been selling its machines in Japan through a subsidiary of Dow Corning, which acts as the company's local distributor. The company has been

satisfied with this relationship. Dow Corning manufactures and
sells silicon raw materials into the marketplace. The firms that
buy raw material from Dow Corning need to slice it and thus are a
potential customer for the company. The size of the market for
slicing machines can be calculated easily from the volume of raw
materials sold. Dow Corning has good access to and knowledge of
the customer and a working sales force in place in Japan.

The company's largest competitor in Japan is a Japanese
firm that presently has a 90 percent market share. This firm
approached the company in the fall of 1984 with a proposal to
form a joint venture. The company declined the offer without
hesitancy. The company feels strongly that joint ventures with
competitors do not work; once technology is transferred it cannot
be taken back, and lawsuits in foreign courts are difficult and
expensive. The company believed the offer was motivated by a
slight drop in the Japanese firm's market share and the recognition that the company had designed superior machinery in terms
of performance, maintenance, and reliability. The Japanese
competitor has since copied the company's pivot-arm technology
and presented this new machine in December 1984, a few months
after the company declined the joint-venture offer.

The company indicated that its current marketing position
in Japan is not satisfactory over the long term and that selling
through Dow is not a viable approach if the company wants to
significantly enlarge its market share in Japan. It is therefore
considering taking the two Japanese employees currently with
Dow in Japan, who are in fact being paid by the company, and
splitting them off into a separate wholly owned subsidiary that
would act initially as a distributor, but ultimately as the market
grew it would evolve into an assembly operation.

The company is not at all attracted by the idea of a joint
venture. It is wary of setting up a full manufacturing operation in
Japan for fear of losing its technology. Given combined goals of
enlarging the company's business while protecting its technology,
a strategy of establishing a wholly owned marketing, distribution,
and assembly operation within the Japanese market will probably
be adopted. It is clear to the company, however, that the
semiconductor industry is shifting toward Japan, and that to be
viable over the long term, the company will have to enlarge its
presence there.

CASE NO. 12

The firm is a publicly held company in the semiconductor
equipment manufacturing industry. The company's two main

product lines for this industry are photolithographic pattern equipment and wafer processing equipment, both of which are used in the semiconductor manufacturing process. The company's machines are recognized by the semiconductor industry as leading-edge products for producing integrated circuit devices. Many of the 256K Dynamic RAMs, the newest generation of very large scale integration (VLSI) devices, are already being manufactured with the company's equipment. Net sales in 1984 were roughly $300 million. Thirty-five percent of total sales are currently generated outside the United States.

The company has been involved in the international markets for over twenty years. The company began doing business in Japan in 1966 through a well-known Japanese trading company. The agreement was for distribution and service; however, the trading company employed another Japanese firm for servicing and installation of the equipment. The equipment is relatively expensive; the photolithographic equipment costs from $700,000 to $1 million per unit, and the wafer processing system costs from $150,000 to $250,000 per unit. In these early days, the company had about a 95 percent share of the free-world market for its product lines. The company had no direct competition in Japan at this time.

In its photolithographic line of equipment, the company had been using optics manufactured by a Japanese company. These were expensive optics costing roughly $35,000 apiece and were sophisticated devices specially designed and manufactured to the company's specifications. In about 1974, the company changed its supply of optics to a non-Japanese source because it believed that the non-Japanese company was offering a higher grade of precision optics. By 1972, the company's share of the world market had dropped somewhat, but was still around an 85 percent share with the majority of sales outside the United States occurring in Japan. The company's source of revenues was divided 55 percent from the United States, 35-37 percent from Japan, and 10-13 percent from Europe.

By 1976-1977, the Japanese firm that previously had supplied the company with optics began to shift its resources into the development of high precision optics and equipment. This product line was essentially based on the company's technology, using the specifications and training provided to the Japanese firm by the company during their years of association. This Japanese firm began to develop a line of products that directly competed with the company's own photolithographic line. This shift in emphasis was heavily supported by MITI, which was helping the Japanese firm acquire technology, supplying it with funds, and perhaps encouraging other Japanese firms to buy this company's products.

The Japanese firm was matched by MITI with two leading Japanese device manufacturers to develop equipment for this market. By 1980, the Japanese company had nearly 15 percent of the Japanese market.

It was after this Japanese firm came into the market with competitive products that the company began to detect changes in attitude of many Japanese customers. There was no question that an attitude of "buy Japanese" was developing. The company invested large amounts of its money to develop new features, which later the Japanese competitor would copy. Although it cannot be specifically proved, the company felt it was being treated unfairly by certain customers.

In 1980, the company decided it needed a direct presence in the Japanese market to preserve its waning market share. Another Japanese firm had come on the market with products that competed with the company's photolithographic line. The company sensed that it was too far removed from its end-user. The company began discussions with its Japanese distributor about forming a joint venture in 1981. The joint venture was established in late 1983, with ownership divided evenly between the two partners. The initial joint-venture arrangement covers installation, service, and some applications work. Sales were not to be put into the joint venture until January 1986. One of the primary reasons given for this is that the Japanese partner said it could not easily transfer personnel from those areas into the joint venture because the joint venture is not wholly owned by the Japanese firm. It had taken considerable negotiation to get the January 1986 commitment, and as a consequence new people were hired in the interim. This has created requirements for additional training. Eventually, the company intends to transfer technology and manufacturing know-how to the joint venture. Some of the research and development for the Japanese market will be a joint effort between American and Japanese engineers employed by the joint venture.

The company is working very hard to make this joint venture successful. Although the partner is not a manufacturer, the company believes its joint venture has good chances for survival. The joint-venture agreement also gave the company substantial control: the company decides which products will be developed and where these products will be developed. It determines when a product will be introduced to the market, and it determines the worldwide pricing. The company believes that the key to success will be in maintaining a good relationship with its partner and properly targeting its R&D efforts to developments in the Japanese marketplace.

The company nonetheless is dubious about the Japanese government's intentions in the marketplace. The company's market share has declined over time. There are occurrences that lead the company to believe there is in the marketplace an influence and a resultant preference for products manufactured by indigenous firms. The company also expressed concern about the recent attempt by MITI and the Japanese semiconductor equipment manufacturers to form an independent Japanese Semiconductor Institute and withdraw from the worldwide Semiconductor Equipment and Materials Institute (SEMI). This new group could have its independent standards group. The company views this as an attempt to exclude American firms from benefitting from the future development of the 1-megabit and 4-megabit devices in Japan.

In Europe, the company sold through a distributor owned by a U.S. company until 1973, when it bought out the distributor and set up its international group in Switzerland. This organization continued to install and service the company's products and performed small amounts of applications work. In 1979, the company began doing some final assembly of one of its products in West Germany. Over the next few years, this operation evolved into a manufacturing facility with some engineering, manufacturing, and assembly taking place. Even parts were purchased locally. Over this period, the company managed to overcome its U.S. competition for the European market because none of them were manufacturing over there. A German firm is now the company's main competition in Europe in one of its lines.

By 1980-1981, the company's European employees began to urge company executives to consider building up the European operation. Although the employees were advocating a wholly owned operation, company executives favored a joint venture where they could share risk and take advantage of local engineering and management talents.

The company then began to speak to potential joint-venture partners. It had discussions with several European companies. These companies were all in the device end of the market and were users of the company's products. The company decided that one French company's approach to business was most similar to its own philosophy; it tended to be more entrepreneurial and open to new ideas. For the most part, the decision was based on a good rapport between the French company and company executives. Ownership in the joint venture was divided evenly between the two firms.

The joint venture in France is a fully integrated facility performing R&D, engineering, manufacturing, sales, marketing, financial, and administrative functions. The company's products are selling well throughout Europe, and market shares have increased since the establishment of the joint venture. Over 10 percent of total sales to the semiconductor market are now generated in Europe. The company has increased its market share; it attributes much of this to the joint-venture activity.

CASE NO. 13

The company is part of a larger publicly traded international electronics firm, founded in the 1940s, that designs, manufactures and markets high-technology systems and components. These products include microwave, power, and special-purpose electron tubes, solid-state components, and equipment for defense and communications; analytical instruments for science and industry; wafer fabrication equipment for the semiconductor industry; radiation equipment for cancer therapy; vacuum equipment and leak detectors for industrial and scientific processes; and miniature magnetic components for computers and telecommunications. The company manufactures equipment and software for the production of integrated circuits; it is the world leader in ion implantation and physical vapor deposition. The company's main products are ion implantation, sputter-coating and lithography equipment; these products are sold around the globe.

The company's experiences selling overseas have been mixed. Within the last five years, the company encountered trouble in Great Britain after it sold a piece of research equipment to a British university. The company's local British competition protested loudly against the university's purchase of non-U.K. equipment. These protests were registered at all levels of the government, including discussions with Prime Minister Thatcher. Needless to say, after this negative publicity the company has not been able to sell another piece of equipment in the United Kingdom; potential customers have been warded off.

Interestingly, the company was recently approached by an individual of British nationality for a license to build the company's equipment in the United Kingdom. Basically, the individual asserted that, although the company could not sell its equipment in the United Kingdom, the company's equipment is superior to that of its British competitor. This individual maintained that he could sell large quantities of the equipment in the United Kingdom if the company would grant him a license to manufacture in the United Kingdom. Though the company currently is unable to make sales, it refused the license on the grounds that the U.K. market was not large enough to risk losing its technology.

The company has been involved in the Japanese market for over twenty years. In 1967, the company established a Japanese joint venture with NEC. NEC owned 51 percent of the joint venture and the company owned 49 percent. At that time non-Japanese companies could not own a controlling interest in a Japanese company. As part of the joint-venture agreement, the company licensed its technology to the joint venture for manufacture in Japan. This technology included vacuum products, coating, metalizing, sputtering, and surface analysis. After about five years, the company began to realize that the joint venture was developing products based on the company's technology that were competing directly with the company's own products. Slowly, the joint venture began to exert pressure on the company to sell its newly developed products in the United States. In addition, the company was dissatisfied with the level of sales of its products in the Japanese market. Unfortunately, the joint venture was structured in such a way that the Japanese partner had considerable leverage over the company and the company was committed to licensing technology as a condition of the joint-venture agreement. The joint venture thus grew in an uncontrolled manner; the company was insufficiently involved in the operations of the joint venture to foresee any of these developments.

When it finally became clear to the company that its joint-venture partner was only after its technology, it stopped the flow of information. Over time, the company has been able to sell off most of its share in the joint venture. Today, the company owns about 19 percent of the joint venture, but for all practical purposes, the two companies have no relationship. This Japanese firm is doing very well in the Japanese market; virtually all of its technology came from the company.

After this experience, the company used a distributor to sell its products in the Japanese market. This arrangement has worked well for the company. The company is the distributor's largest client, and both firms worked hard over the years to make this business relationship work. Over time, Japanese customers began to request more and more special orders. Eventually, it became impossible to service the market effectively from a U.S. manufacturing site. In addition, the Japanese had moved to the forefront of the worldwide semiconductor industry, and it was clear to the company that it had to have a strong position in the Japanese market. Thus the company continued this distributor relationship until it decided that it needed a more substantial presence in the Japanese market.

A joint venture was established in 1984. Ownership is evenly divided between the company and its Japanese partner. The joint venture manufactures, sells, and services the company's product for the Japanese market. The joint-venture agreement was written very carefully to avoid ambiguity. The company closely monitors every drawing that is sent over to Japan and is actively involved in every aspect of the joint venture. The company is working very hard to avoid the problems that had occurred with its previous joint venture.

The company is not unequivocally opposed to transferring technology to the Japanese. For example, the company is currently considering selling some of its lithography technology to the Japanese. It spent $50 million over the last six years developing this technology, but the research has not yet come to fruition. Since the company cannot afford to continue investing its money in this R&D, it is beginning to sell the technology piece-by-piece. Thus far, the Japanese have shown the most interest in acquiring the technology.

The company still voices some reservations about the viability of overseas joint-venture arrangements, in general. The company chose to establish a joint venture because it believed it would be unable to hire good personnel in Japan and it did not have the time to slowly build a work force. Nonetheless, the company views the 100-percent-owned subsidiary as the wave of the future for penetration of the Japanese semiconductor industry and notes that several U.S. firms have already chosen this route.

During its many years of involvement in the Japanese market, the company has experienced varied sales environments. For the most part, when the company offered a product that did not have a technological equivalent in Japan, the product sold well. As Japanese technology began to catch up with the company's product line, Japanese customers demanded more and more special features and product modifications. Since it is economically unfeasible for the company to perform these tasks, the company began to lose sales as its Japanese competitors gained ground.

NOTES

1. United Nations, Center on Transnational Corporations, Transnational Corporations in the International Semiconductor Industry, 1984, p. 151.
2. Ibid., p. 152.
3. Ibid., pp. 455-456.
4. Ibid., p. 152.
5. Ibid., p. 159.
6. Small Business Administration Data Base.

118

7. Organization for Economic Cooperation and Development, The Semiconductor Industry, Trade Related Issues (Paris, 1985), pp. 19-20.

8. Dimitri Ypsilanti, "The Semiconductor Industry," OECD Observer (OECD, January 1985), p. 16.

9. J. K. Paul, ed., High Technology International Trade and Competition (New Jersey: Noyes Publications, 1984), p. 233.

10. Bro Uttal, "Japan's Latest Assault on Chipmaking," Fortune, September 3, 1984, p. 79.

11. Figures from Semiconductor Industry Association.

12. Kathleen Wiegner, "Europe Fights Back," Forbes, August 12, 1985, p. 82.

13. Ibid.

14. U.N., Center on Transnational Corporations, International Semiconductor Industry, p. 249.

15. U.S., Department of Commerce, Bureau of Industrial Economics, French Government Assistance to Its Domestic Computer Industry: Lessons and Implications, no. BIE/IR-83/01 (June 1983), p. 8.

16. "Electronics Newsletter," Electronics, August 19, 1985, p. 12.

17. Bro Uttal, "Japan's Latest Assault on Chipmaking," pp. 78-80.

7
SOFTWARE

The worldwide software market is booming. Growth is occurring at a swift pace from $10.3 billion in 1981 to $18.5 billion in 1983; by 1987, the market is projected at $55 billion.[1] This growth is not surprising, given the strategic importance of the software industry to so many other technology-intensive industries, such as the computer and telecommunications industries. With the growth in computer-aided design and computer-aided manufacturing (CAD/CAM), the strategic importance of the software industry to other industrial sectors of the economy will multiply. Indeed, the importance of the software industry to the future of U.S. industry and technological innovation is difficult to overstate.

The U.S. software industry is made up of a large number of suppliers and markets and a wide variety of software products. The markets range from personal computers referred to as the "low-end," to large or super-large computers referred to as the "high-end." Software products range from integrated systems and custom software to packaged software.

The growth of the U.S. software industry was stimulated by IBM's announcement in 1969 to price software separately from hardware. This event spurred the growth of independent software suppliers and a packaged software market in the United States. The three principal suppliers of software are computer hardware manufacturers, independent software suppliers, and system integrators. The firms in these last two categories tend to be small firms that earn most of their revenues from software, integrated systems, and computer services.

In 1983, there were an estimated 4,850 firms in the U.S. computer software and services industry. Forty-six percent of these companies are in the packaged software and software

119

products business; 29 percent are in the professional services business, which includes custom software and contract programming; and 25 percent are in integrated systems. Of these three sectors, the packaged software and software products group registered the highest rate of entry to the industry by new firms between 1982 and 1983: this sector increased in size by 20 percent, while professional services and integrated systems grew 4 percent and 8 percent, respectively.[2]

Software revenues have increased substantially over the last six to seven years, with large U.S. computer hardware manufacturers accounting for the largest share of revenues. Statistics for 1982 show that large companies (over $10 million in size) comprise only 2 percent of all software products firms, 5 percent of all professional services firms, and 3 percent of all firms in the integrated systems sector. These same firms, however, accounted for 51 percent, 62 percent and 38 percent of total revenues for their respective groups, indicating a high degree of concentration within the U.S. software industry.[3]

Though the U.S. software industry was originally dominated by the large hardware manufacturers, the transformation of the software industry from low-volume, high-priced products for use with large computer systems to high-volume, low-priced products for use with small personal computer systems encouraged the emergence of many independent software companies such as Micro Pro International, Lotus Development Corporation, and Microsoft.

The packaged software sector is, indeed, the most dynamic. As a percentage of total software industry revenues, packaged software increased its share from 52 percent in 1981 to 60 percent in 1983. This is the result of several factors, one of which is the increase in demand that accompanied the boom in the personal computer market. In addition, the packaged software sector is characterized by high initial R&D expenditures, but decreasing marginal costs of production and high-profit margins.

Between 1981 and 1983, worldwide revenues of packaged software by U.S.-based software firms grew at an annual rate of 40 percent. The fastest-growing sector within this market is the low-end, specifically personal computer system, market; revenues grew 83 percent annually between 1981 and 1983. Packaged software for small personal computer systems has been an attractive market because it involves a high-volume product with a relatively short life cycle and low maintenance cost. Companies manufacturing products for personal computers include Visi Corp, Microsoft, Digital Research, and Lotus Development Corporation. The average profit margin for a representative group of these firms was estimated at 23 percent for 1983.[4]

U.S. software producers traditionally have enjoyed a position of technological leadership because of the strong emphasis on research and development in the country and the industry's ability to incorporate R&D breakthroughs into marketable products within a reasonably short period of time. Recent figures show that the United States maintained a 70 percent share of world software markets during the early 1980s. The United States is expected to increase this share to 75 percent by 1987.[5] Total 1982 revenues of the U.S. software industry were ten times larger than its nearest competitors, France and Japan; more than half of these revenues came from packaged software, while most of the leading foreign suppliers gained the bulk of their revenues from custom software.[6]

The European software market, estimated at $10 billion in 1984, is growing at an annual rate of 11 percent.[7] Even though European programming is considered first class, the overwhelming percentage of software used in Europe is American, and the percentage is increasing. Very few European products have made a significant impact on the international markets. Recognition of the importance of the software industry to computer sales and other vital industrial sectors has, however, led European countries to place a greater emphasis on the development of their own domestic software industries. This emphasis is particularly pronounced in countries that have strong domestic computer targeting programs such as France.

France is the leading software supplier in Europe and the second-largest software market in the world. The market, valued at $2.2 billion in 1984, has been growing about 15 percent annually over the last few years. This growth is expected to continue through 1990. Imports of software into France have been growing at an annual rate of about 19 percent, and in 1984 they supplied 24 percent of market demand. U.S. products account for roughly half of total imports.

The French software industry differs significantly from the U.S. industry in that contract programming and consulting activities are the main source of revenue. Cap Gemini Sogeti (CGS), the top French producer, derives all of its revenues from these activities. In fact, France's top-ten producers derive 84 percent of total software revenue from contract programming and consulting.[8] Software products and integrated systems have, however, been the fastest-growing sectors, which indicates that they may build some self-sufficiency in this area in the future. Nonetheless, French producers currently supply less than half the domestic demand for packaged software. For the most part, domestic demand is satisfied by U.S. packaged software products that are sold through local distributors who pay sizeable royalty payments to license these products.

French software firms are making serious efforts to establish themselves abroad. CISI set up its own operation in the United States and in 1983 acquired Wharton Econometrics Forecast Associates Inc. in an effort to diversify into economic information systems. CGS has also been active abroad; in 1983 it derived 25 percent of its total revenues from its four U.S. companies. Nonetheless, an examination of the top French producers shows a definite reliance on the domestic market: CISI derives 80 percent, Societe Generale de Service et de Gestion (SG2) derives 85 percent, and CGS derives 65 percent of total revenues from the French market.[9]

The French software industry, with its roots in the late 1960s, is younger than its American counterpart. This industry grew largely out of the first phase of the Plan Calcul (1967-1971), a government program for the development of the domestic computer industry which included a Software Plan. The Software Plan limited the entry of highly competitive U.S. software firms to the French market and encouraged subsidized hardware manufacturers to contract out computer services and software development to French companies. During the second stage of the Plan Calcul (1972-1975), the French government spent roughly $15 million funding R&D in software development.[10]

The French software industry has also received considerable assistance as a by-product of the Plan Filiere, begun in 1982, which emphasized CAD/CAM, modules for basic mini- and microcomputers, engineering software with artificial intelligence, training and education, and research and development, and through participation in ESPRIT (European Strategic Program for Research and Development in Information Technologies) research projects. As part of the Filiere Electronique, the government funded a research project into computer-aided translation. A prototype system that translates text written in French into nearly faultless English will be demonstrated in September 1985 by five French organizations that have been working under the aegis of this program. This project has been under development since 1983 and administered and funded by the French Agence National de l'Informatique; lead responsibility for the development was taken SG2.[11]

In addition to the assistance received through government programs, the French software industry has benefitted substantially from "buy French" attitudes in both the public and private sector. Furthermore, French end-users are also very demanding; considerable customization is necessary for non-French suppliers to produce a product that is acceptable to the French consumer. Analysts do claim, however, that non-French firms can do well in the market if they make a significant commitment to the market (such as an R&D or manufacturing facility within the country) and present the product in a distinctively French style.

In 1982, the U.K. software industry generated an estimated $700 million in revenues. Half of these revenues came from integrated systems software and the remainder was divided roughly evenly between custom software and packaged software.[12] Many of the United Kingdom's top software producers are subsidiaries of large corporations or organizations, such as British Petroleum, ICL, and the National Coal Board, and benefit considerably through these associations. To date, small firms have not played a major role in the industry.

Technologically, the U.K. software industry is considerably advanced especially in areas of sophisticated application software such as videotex, electronic mail, and transaction switching. University research in the software field is also quite impressive. Unfortunately, British firms seem to lack the marketing skills to translate these strengths into high-volume marketable products. In addition, the British financial system does not adequately support entrepreneurial ventures or start-up enterprises. In this respect, the British Technology Group (BTG), which provides equity and loan finance for innovative ventures, has been helpful. In the computer and software industries, Data Recording Instrument Company, Q1, Insac, and Logica have all received assistance from BTG. BTG will provide 50 percent of the cash flow required by a specific project, which is recoverable as a levy on future sales. In addition, BTG takes equity positions in investment ventures ranging from 10 percent to 100 percent.

In an effort to strengthen its domestic software industry, the British government has implemented programs directed specifically at these weaknesses. One such program is the Software Products Scheme (1981-1985) which has allocated $65 million for direct, nonrecoverable grants for up to a third of development and marketing costs of software products. Support under this scheme is specifically directed toward those products that greatly enhance the productivity of the user, or whose development will be made considerably more effective through the use of productivity aids, or toward those that will have a major effect on the international competitiveness of the developer.

The Alvey Software Program (1983-1988) is another government program in which $114 million will be spent on software research and development with an emphasis on software engineering. Funding for this program is being contributed by the Department of Industry, Department of Education and Science, Ministry of Defense, and British industry.

The West German software industry is dominated by the large computer hardware manufacturers, in particular, Siemens and Nixdorf. Independent software manufacturers have not played a significant role in the industry to date. West German hardware manufacturers as a whole are currently generating 80 percent of total revenues from sales of hardware and 20 percent from sales of software. The current market volume for software in West Germany is estimated at $4 billion with a projected annual growth rate of 15 percent.[13] In 1980, figures showed that West German software manufacturers were generating 21.7 percent of total revenue from software packages.[14] Despite West Germany's strength in computer hardware, it remains largely dependent on U.S. companies for a supply of innovative productivity hardware.

For the last five years, the West German government has actively supported the development of its domestic software industry through a series of targeting programs. Beginning in 1980, the Federal Ministry for Research and Technology (BMFT) began funding a Computer Software Program aimed at upgrading software technology research through the improvement of specifications, operating techniques, quality control, and servicing. The goal of this program is to lower production costs and improve reliability of German software products. This program is progressing in various stages: integration of software/hardware development in microcomputers, funded at DM4.5 million (1980-1983); improved product reliability, funded at DM2.2 million (1982-1983); protected data automation systems for separated offices, funded at DM5.4 million (1982-1984); and computer-aided software production, funded at DM3.5 million (1982-1983).[15]

The German software industry also has benefitted indirectly from government policies targeting the domestic computer industry. In 1977, the German government established a significant network of support programs, which covered twelve financial plans in the three main areas termed crucial to future computer development: industrial research and development, data processing applications, and manpower training. Between 1977 and 1980, funding for these programs reached approximately $650 million. During the 1980s, the German government is funnelling funds into research and development assistance to firms and universities through the Federal Research Ministry of Technology Center in Berlin. Funding for the center reached about $150 million in 1982.[16]

The Japanese software industry accounted for sales of roughly $1.5 billion in 1983, expanding at an annual rate of 22 percent. Nevertheless, even though its domestic software industry has grown into the world's second largest, Japan in general continues to lag well behind the United States in availability and capabilities of software, with a particular scarcity of standard, multiple applications packaged software. In 1983, Japan imported approximately $200 million worth of software, while its software exports were virtually nil. Recognizing this problem and the constraints it places on profitability and plans for future overseas growth, the major Japanese computer manufacturers and independent software houses have greatly expanded their software development efforts in recent years. The Japanese government in turn has adopted a wide variety of policy measures in support of these efforts.

The Japanese computer giants--Fujitsu, NEC and Hitachi--and to a lesser degree, other large firms including the major banks, trading companies, and security houses are the dominant players in the Japanese software industry, generally operating through subsidiaries spun-off from the parent firm. According to MITI, there are some 2,000 independent software firms in Japan. However, approximately 75 percent of these companies serve essentially as subcontractors to the large computer manufacturers. Some 90 percent of Japan's software is custom-made as compared with 35 percent in the United States. Japan's major areas of strength at present include computer graphics, specialized applications such as CAD/CAM, and embedded software. Japan has recently begun subcontracting programming to U.S. firms as well as to other Asian countries, particularly Singapore, Taiwan, India, Korea, and China.

As noted above, government support for software development has been strong and multifaceted. The most important Japanese policies and programs are described briefly below.

Since 1980, software firms have been permitted to set aside up to 50 percent of software income realized in a special tax-free reserve account to cover future software development costs. Also, 20 percent of all year-to-year increases in training costs for software engineers can be taken as a tax credit. A program in effect between 1972 and 1980 allowed up to 2 percent of total software sales to be set up as a tax-free reserve to cover program warranties.

The Information Technology Promotion Agency (IPA) was established in 1970, with the objective of promoting computer usage and software technology. Its activities include commissioning the development of programs, improving the availability of credit to software houses by providing loan guaranties, conducting industry surveys, maintaining a register of general-purpose programs, and the administration of permanent software production R&D facilities and programs. The IPA also arranges loans for software firms through Japan's long-term credit banks. Loans granted either directly or via the IPA Trust Fund totalled $682.7 million over the 1971-1980 period.

Software development has also necessarily accompanied the major government-sponsored computer hardware R&D projects. For example, approximately 80 percent of the funds for phase II of the fourth-generation computer project were allocated to the development of operating system software for fourth-generation systems (e.g., memory management, networking, data base management and high-level languages). Two new MITI-sponsored software R&D projects have been launched this year. The SIGMA (Software Industrialized Generator and Maintenance Aids) project, funded at $100 million over five years, aims to develop an automated system for software production. The Interoperable Databases project, funded at $80 million over eight years, is aimed at enabling data bases with different operating systems to exchange information.

Thus far, intellectual property protection has been the most significant software-related trade issue between the United States and its major trading partners, causing particular friction between the United States and Japan. The United States has unofficially afforded computer programs the status of intellectual property covered by copyright law since 1964; this was made official in 1980. Under these laws, individual programmers are granted exclusive rights to use and license their programs for a period of a lifetime plus fifty years; corporate entities qualify for seventy-five years of protection. The U.S. registration process is simple and all documents are confidential.[17] The United States has favored similar protection of its software overseas.

The United States expressed considerable dissatisfaction last year when MITI proposed a new law on compulsory licensing of software defined as in the national interest of Japan. The Japanese then withdrew the legislation in the face of strong opposition from abroad. On June 6, 1985, the Japanese Diet finally approved legislation amending Japan's copyright law to explicitly include computer programs, ending a long period of heated discussions between the United States and Japan. The Japanese law is similar to U.S. regulations in the terms of licensing powers and the duration of protection; however, there are some significant omissions resulting in ambiguities concerning the copyrightability of proprietary algorithms and languages.[18] Consequently, the area of software copyright protection is very likely to continue as a focus of controversy in Japan and could emerge as a trade barrier as the role of software in data processing innovation continues to increase.

With the exception of intellectual property issues, U.S. companies do not appear to be encountering significant barriers in Japan or Western Europe. At this point, U.S.-written software is well advanced technologically and, given the exigency of productivity in the workplace and the strategic importance of software to other industry sectors, foreign countries cannot afford to keep American software out of their domestic market. Moreover, the majority of U.S. software manufacturers, lacking the in-house know-how and experience, have tended to opt for collaborative research with local companies in order to adapt their products and services to the local market. (Cases No. 15 and 16 provide examples.) As a result, U.S. companies in the software field have gravitated toward joint-venture arrangements in which local companies have benefitted as well. These factors have thus far militated against the development of significant barriers in this industry sector.

CASE NO. 14

The company is a small publicly traded firm that has developed a computer software technology that embodies adaptive or self-organizing capabilities. This software technology is capable of accepting unanalyzed data and modifying itself to detect, store, and recognize identifying features of patterns in the raw data presented to it, without being externally programmed to perform such functions. The software is adaptive in the

TABLE 7.1
Summary of software case studies

Firm No.	Total Assets ($ Mil)	Strategy	Reason
14	1	Delayed entry to Japanese market	Inability to obtain adequate patent protection in Japan
15	200-300	Joint venture with local partner in two European countries and Japan; joint R&D in these countries with R&D also at headquarters	Need to look like a local company; need for understanding of local client; EC procurement policies obscured information for nonlocals
16	150	Distributors around the globe with wholly owned sales subsidiaries in key countries; wholly owned manufacturing and R&D facility in Europe; joint R&D in Japan	Need to show commitment to the British market to sell to nationalized buyers; need to work with a local Japanese company to properly transform product

sense that its subsequent response to inputs is a function of those inputs and of the history of previous data inputs to the system. The method also embodies distributed memory capability and is, thus, relatively invulnerable to damage of individual elements. Because the individual elements of future hardware embodiments of this technology can consist of conventional electronic components, the company believes these embodiments would be capable of being assembled from computer chips. The software also has the ability to generalize, that is, to group together similar input events.

The company's principal development effort to date had focused on the recognition of hand-written characters. In this area, the company has been successful in simulating on a large computer the operation of a system capable of recognizing a large data base of randomly distorted, unconstrained, hand-written numerals with an accuracy in excess of that obtainable by conventional means. The company also has created a feasibility demonstrator that recognizes unconstrained, hand-written characters.

The company believes that there are pattern recognition problems, such as understanding continuous speech, that are not capable of being solved by existing technology because they require real-time responses that are much more rapid than can be achieved with serial computers. The company believes that a high-speed hardware embodiment of the software technology can be developed to function as a parallel processor. Although no such embodiment has been built (as of December 1983), such an application, which would probably be in the form of an integrated circuit, would result in much more rapid real-time processing of complex input patterns. The firm's simulations suggest that adaptive procedures that take minutes to perform on a high-speed serial computer can be accomplished in milliseconds by a parallel processor embodied in such an integrated circuit.

The availability of such an IC embodiment would greatly accelerate the company's design of demonstrators by, in many cases, eliminating the need for slow serial simulations of the software's operation. The development of such demonstrators is a key future step in the further exploitation of the technology and of its licensing to IC manufacturers. Although the company has allocated approximately $400,000 to the development and design of such a chip over the next three years, there are significant questions as to whether this can be accomplished, given the company's time and research constraints. With the company's current capital and net proceeds of a recent stock offering, there should be sufficient funds to satisfy the company's cash requirements, and to afford the expenditures required to operate the business through 1986. The development of new revenue sources through licensing of the module is, therefore, essential to the future of the company.

The company has entered into a licensing agreement with a U.S. hardware manufacturer. This is its only licensing agreement to date. The hardware manufacturer has been granted an exclusive license to use the software and related patents and know-how to develop, make, and sell in Japan certain computer accessories, which may include hardware, firmware, and software. The term of the license is until the patent rights expire and, with respect to the know-how, for seventeen years from the date of disclosure by the company to the U.S. hardware manufacturer. The hardware manufacturer has agreed to pay, during the term of the license, certain earned royalties based on the sale of such devices. To maintain its exclusivity, this firm has agreed to pay minimum annual royalties of $500,000 through June 1985, $1 million through June 1986, and $2 million through June 1987, and each year thereafter, during the term of the license.

Thus, the company's derivation of revenues from this license significantly depends on the hardware manufacturer's ability to market embodiments of the software in Japan. Indeed, sales of this product by the hardware manufacturer in Japan are the major potential source of revenues to the company.

It is in this context that the company's ability to protect its proprietary technology through successful patent filings, is a key to its future. The company has been granted five U.S. patents, four of which relate to the entire information processing method (including process and memory circuits), and to specific representative embodiments of this method. The firm also has been granted thirteen foreign patents, with six pending, in a total of eleven foreign countries. To quote from the company's prospectus, "The company believes that the validity and scope of the issued United States and foreign patents are material to the business of the company." In addition to the risk associated with patent protection, the company faces various potential competitors. The development of new computer architectures embodying adaptive parallel processing has received great publicity as well as resources, particularly in Japan. The successful development of Japanese technology embodying such features would represent a serious threat to the company.

In part due to the complexity of the Japanese writing system and in part due to the Japanese interest in factory, office and home automation, pattern recognition is an important focus of current Japanese R&D. A number of private firms, most prominently Hitachi, Fujitsu, and NEC, are working intensively in this field, with image processing and continuous speech recognition apparently the areas of major interest. These firms and others are also working on advanced parallel processing computer architecture. The Japanese government and NTT are also sponsoring a significant amount of R&D in pattern recognition and parallel processing. Under the Pattern Information Processing System (PIPS) project (1971-1980), in which Fujitsu, Hitachi, NEC, Toshiba, Mitsubishi and others participated, joint R&D was performed on technology for the recognition and processing of pattern information, including characters, pictures, objects, and speech. The fifth-generation project is a notable current example of joint R&D on parallel processing and pattern recognition, with the development of intelligent interface systems, capable of natural language processing, voice processing, and image processing, being a major project goal.

The company's unique technology, its modest capitalization, and tenuous revenue stream make its inability to achieve adequate patent protection in Japan all the more threatening. The company filed for patent protection in Japan in 1975, the same year in which its basic patent was filed in the United States. The U.S. patent was issued in 1976, followed by a number of other patents issued in Europe, South America, and Canada. The Japanese patent, however, has never been issued, and this ten-year delay has been a major brake on its growth, its ability to finance adequate levels of R&D, its ability to produce demonstrations of its technology, and, of course, on its ability to raise outside sources of capital. This is particularly the case since the Japanese market is ideal in many respects for the pattern recognition/language-reading capabilities that the company's technology embodies. The obvious interest of the Japanese government in developing an indigenous pattern-recognition capability, as well as the specious nature of the Japanese explanation for the delay in granting patent protection, has led to the conclusion that the delay represents a deliberate effort to close off the market for the company's product, and simultaneously to build a domestic capability that can challenge the company (and its licensees) in Japan and in other markets.

The period between 1975 and 1982 was characterized by the Japanese patent examiner's assertions of his inability to understand the problem that the technology was designed to solve. In effect, he claimed not to understand the utility or application of the adaptive module. The company continued to respond to requests for further explanation, but to no avail. In 1982, the patent examiner rejected the patent application on the basis that it was nonfunctional, that, in effect, it had no utility. This was hardly credible, given the fact that patents had already been issued in other countries, and products embodying this particular capability had been introduced. The company appealed the decision, and the case went to the trial examiner, who raised some of the same questions. The company offered to demonstrate the system and its utility for the trial examiner, who rejected this offer and declined to hear all oral arguments. The trial examiner also raised, for the first time, the issue of prior art in Japan. At this juncture, two contradictory concerns had been raised: (1) that the capability had no use and (2) that the Japanese were already applying it. In supporting the claim of prior art, the trial examiner produced a patent that was simply a Japanese ripoff of a well-known U.S. patent, the Woodrow patent. This was laughable, a transparent effort to slap a Japanese name on a well-known U.S. patent, which, in fact, the company had already demonstrated in its U.S. patent application was not prior art.

At this juncture, the company was able to demonstrate to the examiner that it had successfully distinguished its technology from the earlier patent, but the examiner indicated he would allow the patent only if the three existing claims against the patent were modified. These claims have become so general as to be emasculated, and the ultimate effect of these modifications is to drastically reduce the amount of protection that the patent, when it is ultimately issued, will provide to the company in the Japanese market. The patent is scheduled to be published in Japan in 1986 in its "emasculated" form. After a period of time, during which protests may be heard, the patent will be issued. Thus, after a ten-year period of frustration, expense, delay, and missed opportunities to exploit this product in Japan--and after a ten-year period during which the Japanese could build their own capabilities in this area--a patent unlikely to afford adequate protection will be issued shortly.

The Japanese market is extremely important to the company. Because of the particular capabilities of the company's technology, and the nature of its licensing agreement with the U.S. hardware manufacturer, the technology has enabled this licensee to develop a kanji reader; the product will enable Japanese computers to communicate with each other in Japanese, a capability that could be extraordinarily powerful in the Japanese market. It is uncertain whether the Japanese clearly understood the implications of this capability in dealing with the patent application, although this was apparent. It will, however, be seen as very threatening in Japan when the licensee introduces its product. This will be the case not only because of the pattern recognition capability but also because this system involves parallel processing, one of the key features of MITI's fifth-generation project. In effect, the company has developed jointly with its licensee a product that is a key to the future of Japanese computing.

Now that Japan appears on the verge of issuing the patent, the company's principal concern is that the claims have been so modified as to be nonspecific, and that, as a result, patent infringements will be very difficult to recognize and prosecute. For this reason, the company filed an additional, more specific patent application in 1981, and is about to file another. The company has also attempted to link patents to know-how in its licensing agreement with the U.S. hardware manufacturer. The company expects greater problems with the 1981 patent filing and the one it is about to make, since by this time Japan will most likely have realized the full significance of this technology.

The company has not performed any close studies of the potential market in Japan. However, given the 700,000 or so computers sold in Japan, and assuming a market penetration of 10 percent, the retail value of the licensee's sales would probably be in the $400 million range. The company obtains a royalty on these sales that would total between $12 million and $15 million. It is about a $1 million a year business now, with about three-quarters of these revenues coming from fees paid by other companies for the development of products, and the rest from royalties on sales of products. The company's hopes are for $10 million a year in revenues just on the strength of the Japanese market that its licensee hopes to develop. These numbers convey the tremendous importance of the Japanese market for the company. Nonetheless, the company has made no direct approaches to the federal government.

CASE NO. 15

The firm has been in business for more than ten years. The company provides rather sophisticated computer software systems for a relatively specific market sector. The company is a leader in this market sector in the United States. The company has been quite successful financially.

The company attributes its success to two main factors: its concepts of software design and its in-depth knowledge of the inner structure of its target market and clients. The company has placed particular emphasis on the cultivation and development of expertise in these two areas and on the maintenance of steady, controlled growth. It was the latter of these two considerations that led the company to approach its expansion into the international markets with a great deal of planning and research.

The company began to explore overseas markets in the late 1970s, focusing its attention on Western Europe and Japan. In Western Europe, the company concluded that given the nature of the company's product, the United Kingdom would be a very important market to break into because specific characteristics of the target market in the United Kingdom are vastly dissimilar to those of the target market in the United States. The company knew that it had already proved its ability to deliver comprehensive computing systems and services to its target market in the United States and believed that a success in the British marketplace would be the best way to demonstrate the company's ability to perform in varied environments and to satisfy a wide range of potential customers.

Nevertheless, the company decided that the best way to start its internationalization was to begin in a small country that had a target market similar to the U.S. target market and use it as a pilot. For this initial foray, the company chose Holland. Although it was clear that the product would have to be translated into Dutch to meet end-user requirements, the English language could be used in the business environment. In addition, the target market in Holland was found to be similar to that in the United States, the financial commitment was assessed as modest, and it was felt that if the pilot turned out to be unsuccessful, the exposure in terms of reputation would be limited. The company chose to license its technology to a Dutch firm that was in a related business but was looking to branch out into new technologies. The pilot was very successful and the company felt it was ready to move on to the U.K. market.

For the United Kingdom, the company initially decided on a joint venture because, among other things, it was particularly interested in projecting a British image. As its partner, the company chose the subsidiary of a very large British company involved in the target market that had certain experience similar to that of the company. The company owned 49 percent of the joint venture with the remainder owned by the British partner. The joint venture had a very difficult time selling during the first years. In 1982, the British partner made a corporate decision to pull out of the joint venture. The company decided to continue, however, and was able to buy on attractive terms the remaining 51 percent of the venture from the British parent company.

In late 1982, the British subsidiary won a very large multiple system contract in competition with a large U.K. company. In 1983, the company found a first-class manager to lead the development of the organization. In early 1984, the British subsidiary won its second major contract. More recently, this subsidiary won two more very large contracts. The company's British organization is now profitable and viewed as a real success story in Britain.

The company believes its success in Britain is attributable to several different factors. Within this time period, the European Community passed new regulations governing public-sector procurement of services. The company found that these regulations significantly opened up the British market. Before these regulations were passed, the company had difficulty learning about upcoming tenders and in getting an equal opportunity in bid evaluation. In addition, the company's emphasis on maintaining a very British image helped to make it a success. The company chose a British name for the joint venture, hired only British personnel, used a British sales approach, and provided systems that were as British or more British than systems from other companies. Most importantly, the company had a great deal of patience and determination which eventually paid off.

Initially, the software development for the United Kingdom took place exclusively in the United Kingdom. This development work, however, was based from the beginning on universal principles that would allow the basic software to approach a truly international product. For example, a translation capability was incorporated.

The company's approach to the Japanese market was similar to its approach to the British market. The company spent a lot of time and effort locating the right partner for a Japanese joint venture. In 1982, the company chose a service bureau that was a subsidiary of a large and prestigious Japanese organization.

The Japanese joint venture received three sizeable contracts, each for a five- to seven-year effort. Providing the software and developing the organization to service these contracts are proving very challenging. The company currently has a sizeable group of people working on the Japanese product in the United States and is delivering the product step by step.

The company's product has been conceived to be far superior to any product offered by its Japanese competitors for the target market. Unlike many of these Japanese systems, the company's product integrates a wider range of activities in the target market, rather than focusing on just one of the activities. Also, the Japanese have not as yet shown themselves to be as proficient in the development of software of the type produced by the company. Accordingly, the company hopes to have a significant competitive advantage. The company's product for Japan is Kanji-based; that is, it incorporates the Japanese character set. In line with the company's concept of international software development, this Kanji capability was built into the basic architecture of all the company's software so that any system installed anywhere in the world has this capability as a profile option. Hence, the company is slowly developing a universal product that can be sold anywhere in the world.

The company hopes through its efforts to evolve into a truly international enterprise over the next ten years. In addition to the internationalization of its products, accomplished through its central development group, the company has begun to actively recruit personnel from universities around the world, including Japan, West Germany, Holland, and Ireland to work at the company's U.S. headquarters.

Although the company has experienced some difficult times during this period of internationalization, the results of its efforts have thus far been impressive.

CASE NO. 16

The company is a publicly traded firm specializing in business software for microcomputers. As a corporate goal, the company has emphasized innovative solutions to business problems by joining technologically advanced software products with an organization capable of successful delivery, strong product support and financial independence. The company has a strong industrywide reputation for producing quality products, and for its programs for selling, marketing, and supporting those products; this reputation has set new industry standards. The company currently employs about 1,000 people; net revenues this past year were over $150 million. Until 1984, there were no significant revenues generated outside the United States.

The company did not attempt to make any serious efforts to venture into overseas markets until January 1984. Initial efforts were directed toward the European market. At that point the company was not doing any business in Europe. After much research, the approach adopted by the company was to establish local distributors in each European country on an exclusive basis, a wholly owned marketing support organization in France, West Germany and Great Britain, a software research and development group in Great Britain, and a manufacturing facility in Ireland.

In each country in Western Europe, the company identified a distributor partner. It then trained each distributor's personnel and co-funded a wide range of marketing programs including dealer training, advertising promotions, and the establishment of product support lines in each country. In the beginning, the company helped to launch the product line in each country and sent its own personnel to help the distributor's salesmen call on customers. The company had its products translated into local languages as quickly as possible and provided all marketing materials in local language. The company also developed advertising campaigns for each market that took into consideration the particular market needs and cultural values of each country.

Although the company has now set up its own wholly owned sales and marketing support operations in France, West Germany, and Great Britain, it still maintains its relationships with its original sole distributors in those countries. The company's sales and marketing support operations there call on the local dealers and market to corporate accounts; each sales and marketing support operation has its own administrative and financial functions.

The R&D facility located in Britain plays an important role in the company's overall marketing strategy in Europe. It gives the company the ability to react instantaneously to software problems or opportunities to adapt the company's software to new developments in indigenous personal computers or peripherals. The company believes that the European development operation, in addition to enabling the company to react quickly to market needs, also helps the company's marketing and sales operation there to be perceived as a real commitment to the British market. This helps the company to sell to the British government and local banks and oil companies that have strong nationalistic tendencies. The company chose the United Kingdom for the location of its development group for two main reasons: the absence of a language barrier; and the strong supply of local, technically qualified individuals.

The company chose to set up a manufacturing facility in Ireland so that it could control the manufacturing process and the distribution of its products very lightly and obtain some tax advantages. The key determinant here was the perceived need for operational control. None of the company's competitors has done this; the company believes that all these reasons explain why it has a dominant position in every market it has entered in Europe.

It is still too early to assess the success or failure of the company's strategy in Europe. At this point it appears to be working very well. The company's products have been selling well there and do not appear to be encountering any indigenous barriers. At the end of 1984, after only six months in the European market, sales in that area were running roughly 10 percent of total sales.

The strategy developed for the Japanese market had three distinct stages. The first stage involved finding a Japanese software house with which to undertake the process of modifying the company's existing software for use in the Japanese market. The second stage involved finding a local Japanese distributor to help the company break into the Japanese market. The third stage involved building a wholly owned sales and marketing support operation in Japan to slowly take over the distributor function, and a software development group within the company that could slowly take over the R&D function. The prime goal of this strategy is to build a name very rapidly, provide a high level of support and service, demonstrate true value-added to customers, and have very close relationships with major accounts to effectively lock out potential competitors. The company will then maintain that lead by constantly upgrading and releasing new versions of its software.

The company first visited Japan in November 1984. The purpose of this visit was to assess market opportunities and to begin the process of identifying a local software development partner. The company believed that to do well in the Japanese market, it would have to offer more than a translated product. Basically, the company would have to rebuild its products based on and tailored to the needs of the Japanese businessmen. The company did not have this kind of expertise in-house, hence, the need to find a Japanese software development partner. After several more trips to Japan, the company located a Japanese software house that it believed could be an effective development partner. This is a small Japanese software firm that, until recently, was involved primarily in mainframe software. The firm presently markets one of the best-selling word processing softwares in Japan. The company believed that this word processing capability would be a great advantage in the translation of the company's product to Japanese. This could afford the company's product a marked competitive edge because most of the comparable products on the market simply have a few Kanji characters plugged into the menu; no real attempt has been made to address Japanese user issues. The company believes that if it could do this effectively, the product could be a true success.

Choosing a development partner was not easy; the company consulted beforehand with many individuals throughout Japan. Mainly, the company was looking for a Japanese company that had the right technology and developed expertise, could deal effectively with a U.S. partner, and had the type of management personality that would blend well with the company's own corporate style. Everyone the company talked to was extremely open and willing to help the company locate the right partner. Basically, the company believes that doors were opened for the company because Japan needs a good, high-performance, productivity software. The company's software could provide a needed shot in the arm to the Japanese personal computer industry that would allow it to sustain rapid growth over the next five to ten years. In addition, some of the Japanese hardware companies interested in the U.S. and European markets view the company as a possible entree to those markets, and thus, are willing to help the company gain access to the Japanese market.

The agreement between the company and its software development partner includes provisions to license in perpetuity the company's technology for specific applications chosen by the company. The two firms had open and frank discussion about each firm's intentions and the alternatives available to each firm in the event that either firm wishes to enter the other's market. The company's decision to license its technology was positively influenced by the Japanese government's decision to grant copyright protection to computer software; the company viewed this as an indication of the country's willingness to cooperate with foreign software businesses seeking to enter the Japanese market.

The company plans to set up a wholly owned subsidiary in Japan for sales and marketing support, but will start out with a third-party Japanese distributor to help it break into the market. The company will, however, be responsible for all marketing from the start. Over time, the company's wholly owned sales and marketing subsidiary will begin to assume more and more responsibility for the distribution function. The company will probably, however, always have an independent distributor in Japan.

The company expects there to be a rapid increase in demand in 1986 for productivity-based quantitative analytical tools, such as the company's product, but that demand will level off if the proper marketing programs are not in place. The company views very large corporations as its main target market in Japan. Hence, the company strives to provide easy-to-use, easy-to-learn products and to have the right training and educational program in place. The company's main competitors in Japan are now and will continue for the foreseeable future to be other American companies. The business software market is fragmented with 300-plus independent software vendors in Japan competing for a reasonably large market. None of these companies has yet to make a major impact on the marketplace.

The company is taking a long-term view of the opportunities in Japan and is prepared to invest substantially up-front to achieve its goals. Its interest over the long run is to have a major presence in Japan in terms not only of market penetration but also in establishing links with Japanese companies and technology. The company foresees that the latter of these two interests could bear fruit for the company on a global basis. There is certain technology that is state of the art in Japan, such as printers, high-resolution displays, flat panel displays, and graphics, that could open up avenues for the company to develop even more creative, rather than productive, software for different market opportunities.

NOTES

1. U.S., Department of Commerce, International Trade Administration, A Competitive Assessment of the U.S. Software Industry (December 1984), p. 33.

2. INPUT data, provided by ADAPSO.

3. Department of Commerce, Competitive Assessment of U.S. Software, p. 6.

4. Ibid., p. 13.

5. Ibid., pp. 32-33.

6. Ibid., pp. 35-36.

7. Alan Cane, "U.S. Imports Dominate the Market," Financial Times, June 24, 1985, p. 6.

8. Department of Commerce, Competitive Assessment of U.S. Software, p. 39.

9. Logica, "Crowning Europe's DP Royalty," Datamation, September 1, 1984, pp. 157-160.

10. U.S., Department of Commerce, Bureau of Industrial Economics, French Government Assistance to Its Domestic Computer Industry: Lessons and Implications, no. BIE-1R-83/01 (June 1983), pp. 4-5.

11. "French Ready Translation System," Electronics, July 1, 1985, p. 27.

12. Department of Commerce, Competitive Assessment of U.S. Software, pp. 42-45.

13. Karen Williamowski, "CMP Industry Sector Analysis Form," July 17, 1984, pp. 1-7.

14. Organization for Economic Cooperation and Development, Committee for Information, Computer and Communications Policy, "Software: A New Industry" (draft) (Paris: OECD, February 2, 1984), pp. 85-86.

15. The Futures Group, "Data Base on Industrial Targeting," 1984.

16. J. K. Paul, ed., High Technology International Trade and Competition (New Jersey: Noyes Publications, 1984), p. 72.

17. Jon Choy, "Tokyo Expands Copyright Law to Cover Software," JEI Report No. 28 B, July 19, 1985, p. 9.

18. Ibid., pp. 9-10.

8
TELECOMMUNICATIONS EQUIPMENT

The worldwide telecommunications industry has undergone significant change over the past two years, which is highlighted by the privatization of British Telecom and Nippon Telephone and Telegraph, both of which were previously state-owned monopolies. This restructuring signals the opening of new markets and provides increased opportunities for global competition. These factors are certain to lead to a proliferation of new products over the next few years. Currently, the United States continues to lead the worldwide telecommunications industry in terms of technological innovation due to the size of and long-standing competitiveness within its domestic market.

By 1993, the worldwide market for telecommunications is projected to surpass the trillion-dollar mark. Today, the United States accounts for 40 percent, or $142 billion, of worldwide telecommunications expenditures, which collectively amount to about $335 billion. With sales of nearly $120 billion annually, transmission services is the largest U.S. communications market; the growth rate in this sector is 12-13 percent annually. The communications equipment market, which includes central office data communications, radio, and customer premises technologies, has sales of about $20 billion annually; the PBX is the principal product in this market segment. The fastest-growing segment of the telecommunications marketplace is data communications with an annual growth rate of 30 percent. Within this $2.5 billion market, modems represent the largest component with annual sales of $850 million.[1]

The U.S. telecommunications equipment industry historically has been dominated by a small number of very large producers. The largest, Western Electric, accounts for roughly 65 percent of total U.S. output. Western Electric is the manufacturing arm of American Telephone and Telegraph (AT&T), which is the largest consumer of telecommunications equipment. In 1982, AT&T, through its Bell System, serviced approximately 80 percent of the U.S. telephone network. Traditionally, both the Bell System and GTE largely depended on their captive suppliers. GTE services approximately 10 percent of the U.S. telephone network and accounts for approximately 10 percent of U.S. telecommunications equipment sales. Although equal access has not yet become a reality, GTE, the industry's No. 3 carrier, boosted its long-distance revenues 70 percent, to $1.2 billion in 1984.[2]

The remainder of the U.S. industry is composed of independent equipment manufacturers and includes companies such as ITT, Collins Radio, TRW Vidar, Rolm and TIE, as well as a large group of smaller specialty companies. ITT is the second-largest producer of telecommunications equipment in the world, with the majority of its sales and manufacturing in Western Europe. While ITT does offer a broad line of telecommunications equipment, this is atypical of the companies in the independent equipment manufacturers group, which tend to focus on narrower product areas such as transmission or switching equipment.[3]

Historically, the nonaffiliated or independent equipment manufacturers have sold principally to independent telephone companies and private buyers. In 1982, independent telephone companies accounted for 18 percent of demand. Interconnect companies, specialized common carriers, domestic satellite firms, and value-added carriers constitute a very dynamic group of equipment purchasers whose demand for products is on the rise, a large number of which are small businesses; this increase will tend to favor nonaffiliated equipment manufacturers. Between 1976 and 1982, the number of firms in the telephone equipment industry (SIC 3661) increased 44 percent from 271 to 390; small businesses (under 500 employees), as a percentage of this total, decreased from 97.7 percent in 1976 to 95.6 percent in 1982, according to the Small Business Administration Data Base. In 1976, small businesses accounted for 63 percent of total sales and 61.6 percent of total employment in the industry; by 1982, these figures had dropped dramatically to 3.9 percent and 3.3 percent, respectively. In terms of productivity, small firms are outproducing large firms; in 1982, average firms' sales per employee in firms with under 500 employees was $71,000 versus $49,800 in firms with over 500 employees.[4]

Though the United States has long enjoyed a position of leadership in the telecommunications industry, between 1978 and 1983 the American position in telecommunications trade weakened. A continuing increase in imports caused stagnation and, in 1982 and 1983, a deterioration of the U.S. surplus from $159 million in 1981 to $90 million and $60 million in 1982 and 1983, respectively. Though exports grew 13-18 percent per year over this period, imports outpaced this growth by 6-17 percent.[5]

The Japanese telecommunications market is roughly one-quarter the size of the U.S. market. This market is, for the most part, dominated by four Japanese companies: Nippon Electric Company (NEC), Oki Electric, Fujitsu and Hitachi. In 1984, these four firms supplied roughly 70 percent of the domestic market. NEC, with a market share of 37 percent, is the largest supplier to the Japanese market; the other three firms supply 6 percent, 13 percent and 12 percent, respectively. An examination of trade figures from 1978 and 1981 for this sector shows a 17 percent annual growth rate in exports, indicating that the Japanese industry has been a healthy exporter. This growth rate is expected to continue through 1986.[6]

In the fall of 1984, NTT began work on its much publicized Information Network System (INS), a proposed nationwide network of fiber-optic cables and satellites that will provide Japan with a broad range of digital voice, data, and video communications by the year 2000. NTT estimates it will spend between $106 billion and $128 billion to complete the INS, with a first-year budget allocation of $7.3 billion. The demand for equipment that essentially plugs into the INS, including sophisticated digital switches and exchanges, local networks, and terminals, will total as much as $300 billion according to Nomura Securities. Since NTT is not an equipment manufacturer but rather purchases and installs new equipment, enormous sales opportunities for telecom producers are expected to be generated in Japan over the next two decades. The degree of access that foreign producers will have to this market remains unclear.

The telecommunications industry has in recent years been a major focus of U.S.-Japan trade friction. Until 1981, foreign suppliers were precluded from selling to NTT. Although NTT procurement of foreign products has doubled annually since that time, in 1983 U.S. firm equipment sales accounted for only about $140 million, or less than 5 percent of the annual NTT equipment procurement budget of $3 billion. At the same time, taking advantage of industry deregulation, Japanese firms increased their exports of telecommunications products to the United States to about $2.1 billion.

A three-year procurement agreement, renewed in January 1984 by the Japanese and U.S. governments, commits NTT to treat U.S. and Japanese suppliers on an equal basis. Nontariff barriers such as NTT requirements for exhaustive testing procedures, inflexible product design (as opposed to performance) specifications, lengthy design reports, and Japanese-language-only bid applications have been greatly alleviated under the 1984 procurement pact.

Nevertheless, the traditionally close relationship between NTT and its family of supplier firms will not be altered quickly. In 1984, 80 percent of NTT's equipment purchases came from just ten suppliers, and about 40 percent came from just two firms, NEC and Fujitsu. These supplier firms also benefit from joint R&D with NTT, which has several top-level electronics research facilities, and direct transfer of design and manufacturing technology from NTT. NTT also is known to pay its major suppliers quite generously, in some cases, two or three times the market rate. Thus, NTT has traditionally played an important role as a channel for subsidies and technology to several of Japan's major telecommunications firms.

The Japanese telecommunications industry underwent a dramatic restructuring in April 1985, however, with the adoption of a new telecommunications law, which effectively deregulated the industry. The bill transformed Nippon Telephone and Telegraph, previously a state-run monopoly, into a private firm, although the government will sell off its NTT shareholdings only gradually. It also opened the door for increased competition for NTT and Japan's international telephone service provider--KDD. Other firms can now sell telephone and data communications services in Japan, and producers of telecommunications equipment are for the first time permitted to sell directly to consumers rather than only through NTT. It will likely continue to be difficult for new entrants, either Japanese or foreign, to make substantial inroads in sales to NTT. In the area of communications satellites, Japan presently maintains a strict "buy Japan" policy for these systems, refusing to buy U.S. satellites since it desires to build up the capabilities of its domestic industry in this area.

Several U.S. firms are negotiating joint ventures with Japanese partners to provide value-added networks (VANs) and other telecom services. On September 12, 1985, International Business Machine Corporation (IBM) and NTT signed an agreement to form a joint venture to build complex computer networks in Japan that would enable companies to tap into remotely located computers and to exchange text, computerized images, and graphics over telephone lines. IBM Japan will provide the software and computer terminals used in the systems, and will be responsible for marketing the system.[7] This is the first major cooperative agreement NTT has concluded with an American multinational corporation. In an effort to demonstrate its willingness to open its telecommunication market, NTT is reportedly negotiating the purchase of additional equipment from American

Telephone and Telegraph (AT&T) and Northern Telecom. In addition, a contract to provide digital switching equipment was awarded by the Japanese telephone company, Daini-Denden, in September 1985 to DSC Communications Corporation. DSC Corporation was the first U.S. company to be awarded such a contract in Japan.[8]

However, the new NTT will likely remain the dominant force in the market. With more than $20 billion in annual sales, 320,000 employees, and an annual capital budget of about $8 billion, it is Japan's largest company and will continue for some time to maintain its close relationship with the GOJ's Ministry of Posts and Telecommunications, as well as the major Japanese telecom equipment producers.

Taken as a whole, the European telecommunications market is about the same size as the U.S. market. European R&D in telecommunications is of equal quality to that of the United States though not as broad in scope or large in scale, and the quality of European telecom products is reported to be quite high. The European telecommunications industry operates in a market that accounts for one-third of the world's telecommunications equipment sales, and many of the companies do substantial overseas business and are well into the development of sophisticated products such as switches for the integrated services digital networks (ISDN). In 1984, EEC countries recorded a $1.5 billion positive balance against the United States' $300 million deficit in public telecommunications systems and equipment.[9]

Most of the telecommunications equipment markets in Europe fall into one of the following three categories: essentially closed with one or two large domestic manufacturers providing equipment (Germany, France); in the process of liberalization, showing less concentrated national production and greater import penetration (United Kingdom); and smaller, fairly open with a major exporter firm (Sweden) or reliant on foreign sources for equipment supply (Ireland, Austria).[10] The European telecommunications market is characterized by separate national markets in which different technical standards and homologation procedures ensure a de facto protectionism. As a result, nondomestic suppliers have long had difficulty selling equipment for telephone networks in Europe.

The French telecommunications market is dominated by a state-owned monopoly. All service within the French telecommunications network is provided by the Ministry of Postes and Telecommunications (PTT). Responsibility for French telecommunication policies rests with the Direction Generale des Telecommunications (DGT, a branch of the PTT). The DGT determines technical standards and systems needs, establishes domestic and international policies, and performs research and development. The PTT operates the telephone network and buys its products from various suppliers. General PTT policy is to have at least two suppliers for each type of equipment. These suppliers are selected by tender on a competitive basis. The major French suppliers are CIT-Alcatel and Thomson-CSF, both nationalized firms; these two firms supply most of the PTT's requirements.[11]

CIT-Alcatel is a subsidiary of Compagnie Generale d'Electricite (CGE). In 1985, the company generated about $1.6 billion in sales. Approximately half of this business was in France. Since growth in the French market is expected to level off with the completion of a digital network within the next two years, CIT-Alcatel is gearing up to more actively pursue foreign markets. Currently, CIT-Alcatel is involved in forty-four countries around the world.[12]

The French government has spent between $4 billion and $5 billion annually modernizing its telecommunications network over the last nine years. Expenditures are expected to continue at these levels for the next two years. These funds are being used to complete a nationwide integrated digital network. French authorities hope to connect the present system of networks to a single integrated service digital network of nationwide coverage in the near future.

The French telecommunications industry has long benefitted from the preferential treatment afforded French suppliers by the PTT. In addition, the PTT has spearheaded various R&D efforts including Teletel, videotex in a general information retrieval/transaction-processing role, and the electronic telephone directory. The electronic telephone directory moved out of the trial stage in 1983; though the original plan envisioned the establishment of 34 million terminals by 1992, it is expected that only 600,000 will be installed by the end of 1986.[13]

The French telecommunications industry has benefitted indirectly from federally funded development projects aimed at other industrial sectors. In the fourth stage of the Plan Calcul, for example, funds were used to launch the French telecommunications satellite, Telecom I, and for development of domestic data transmission networks. The Integrated Circuits Plan, begun in 1978, directed funds specifically at the development of integrated circuits for telecom applications.

The organizational structure of telecommunications service in the United Kingdom has changed significantly over the last few years. Until 1980 all telecommunications services in Britain were provided by the British Post Office monopoly. In that year the telecommunications functions were split away from the Post Office and vested in a separate government-owned company,

British Telecom (BT). British Telecom was established by an Act of Parliament as a public corporation. In 1985, 51 percent of British Telecom was, however, sold through a public offering for $1.62 billion.

The standards for telecommunications apparatus are written by the British Standards Institution (BSI) and approved by the government. The BSI is an independent organization that functions as the United Kingdom member body of the International Standards Organization (ISO) and the European Committee for Standardization (CEN). Once standards are written and come into force, apparatus that has not been approved previously, including equipment supplied by BT, has to be submitted for type-approval to the British Approvals Board for Telecommunications (BABT). BABT, which charges for its services, arranges for testing of samples to check for conformance with established standards. BABT also examines manufacturing and quality assurance procedures in the factory where the apparatus is produced and follows up with a spot check of facilities about once a year.[14]

In the past, procurement by the British Post Office favored British equipment producers. General Electric Company (GEC), Standard Telephone, and Plessey control over 90 percent of the U.K. telephone equipment market. BT, however, has more latitude in procuring its equipment. This factor should change competitiveness within the U.K. market, resulting in increased opportunities for foreign vendors. British Telecom, with annual sales of $9 billion, emerged from the government deregulation program as an aggressive international competitor.[15] As a private company with a mandate to make profits, British Telecom can no longer afford to show preference to U.K. suppliers. British Telecom now buys exchanges from both U.K. and overseas suppliers; BT recently bought equipment from AT&T and Philips for a new toll-free service.

In West Germany, telecommunications are controlled directly by the State through the Deutsche Bundespost (DBP). The DBP has exclusive rights to provide telecommunications services in the Federal Republic of Germany and is obligated by law to provide all of its services for the general public. Public telecommunications networks are designed, built, and operated by the DBP. All terminal equipment, and its connection to public network, requires the DBP's approval. A substantial proportion of the construction work is contracted by the DBP to private firms. In addition, all products related to telecommunications are produced by private firms. The DBP calls for tenders based on DBP procurement specifications. Type approval is required for practically all types of equipment used in or connected with the public telecommunications network.[16]

West Germany exports a significantly greater share of its telecommunications production than either the United States or Japan. The main supplier is Siemens, Europe's leader in telecommunications, whose activities include communications terminals, private special-purpose and public communications systems, and safety and security systems. Figures for 1984 show that Siemens ranked third in world market share behind AT&T and ITT. Siemens is an export-oriented company; it currently exports 85 percent of its most advanced digital switching systems to sixteen countries with the remaining 15 percent purchased by the German Bundespost. Siemens expects to deliver the first ISDN switching center in Mannheim in 1986; this will make the Bundespost the world's first PTT to operate an ISDN switching center that conforms to CCITT standards.[17]

U.S. manufacturers have encountered considerable difficulty entering the West German market as a result of preferential government procurement policies, and standards and certification procedures. Of particular concern to U.S. companies is the regulation of leased lines and control of transborder data flows by the German PTT. The German PTT charges considerable fees for use of these transmission lines and has strict regulations governing the form in which data can be transmitted across its borders.[18]

Though the European telecommunications industry is competitive both technically and commercially, there is a growing awareness among European policymakers that Europe will fall behind the United States and Japan in telecommunications if it does not make a serious effort to stay abreast in the world market. Europe's position in the world market is eroding: its share of world exports has been declining on average by 1 percent a year during the 1980s. Sales of high-end, new technology products as a share of total sales have not been as strong as the sale of traditional equipment and spare parts.

Given the pace of Japan's advances in electronics and plans for its telecommunications sector and the increased competition in and vitalization of the U.S. telecom market following deregulation, the declining trend in European telecom could accelerate. The protectionist government monopolies that characterize the European market have tended to impede innovation and discourage the development of economies of scales within the European marketplaces. Though Japan spent $1.6 billion to develop two types of central switches and North America spent $3.5 billion for four such systems, Europe spent $6.7 billion for the development of nine types of switches.

It is in response to these concerns that the European Community recently proposed the RACE (Research and Development in Advanced Communications Technologies for Europe) project. RACE is an ambitious research and development program designed to establish common integrated broadband communications (IBC) networks by 1995. This program will lay the groundwork for Europe-wide compatible IBCs that will include the present network structures such as public switched telephone networks, integrated services digital networks, cable-TV systems, and mobile communication, as well as more sophisticated services such as videoconferencing and high-definition TV.

For the definition phase of RACE (1985-1986), $15 million has been committed both by the EC and industry to finance the start of eleven projects. During this phase, resources will be used to develop an IBC network reference model with a suitable architecture, define the IBC terminal environment with a view toward cost-effective equipment, and assess future applications. In addition, long lead-time R&D projects considered essential to communications-related technologies will be undertaken.[19]

Following the definition phase, the first phase (1987-1992) will develop the technology base for the IBC network and help CEPT (Conference Europeene des Postes et Telecommunications) and CCITT (Comite Consultatif International Telephonique et Telegraphique) in formulating specifications and standards. The second phase (1992-1997) is to develop the base for enhanced IBC systems and services beyond 1995. Thus far, exact financial plans for Phases I and II have not been made.[20]

The majority of European firms support the establishment of RACE. U.K. manufacturers are reportedly the most enthusiastic supporters: their telecommunications market is the first to be liberalized and the ensuing rise in competition is beginning to be felt. The RACE program is already seeking to limit the number of main exchange switching systems that British Telecom can buy from foreign supporters. It also is hoped that RACE will boost the U.K. industry's weak export base.

This enthusiasm, however, is not shared throughout Europe. In Germany, Siemens is already well advanced in developing narrow-band ISDN systems and is hoping to get this equpment on the market by 1988. If these developments proceed according to schedule, Siemens plans to have the broadband ISDN network equipment on the market by 1990 or 1992, in advance of the RACE timetable. Thus, Siemens claims that RACE will tie up valuable development resources and aid only Europe's weaker companies. French officials have been even less receptive, claiming that participating countries do not have sufficient confidence in the program to funnel all of their development resources into the EC program which would be required if the program were to be a success. Thus, the future of the RACE project is still unclear.[21]

The strategic importance of advanced communications technology and strong telecommunications networks to national security has led governments to protect their domestic telecommunications industries. The existence of monopolistic PTTs in Western Europe and Japan has provided governments with effective channels through which to implement these protectionist policies. As a result, the widest range of complaints was found in this industry sector, including incompatible standards for equipment (CCITT standards in Europe versus U.S. standards), the use of standards to discriminate against U.S. products, preferential government procurement, technology transfer as a prerequisite for contract award, and issues concerning the exportation of sensitive data.

Almost all the companies interviewed cited difficulties selling to either NTT in Japan or the PTTs in France and West Germany. Companies encountered a range of impediments including outright preference for local products (Case No. 20), less obvious discriminatory treatment (Cases No. 17, 19, 21, 22, and 23), and problems with standards and product certification (Case No. 19). In some cases (Cases No. 21, 22, and 23), the German Bundespost exerted pressure on the U.S. company to transfer technology to a local company through licensing or joint development work.

The problem of selling to markets that require a product to meet CCITT standards was also raised by several companies. Companies have had to produce to local standards, buy local interface equipment, or try to convince local certifications bodies that their product conformed to CCITT standards. Each of these options cost the U.S. company considerable time and money. However, as the large European telecommunications manufacturers such as Siemens, Olivetti and CIT-Alcatel strive to achieve greater economies of scale by expanding their markets beyond national boundaries and competing overseas, there is increasing discussion of the need for a compatible system of standards. As a result, as internationalization of the industry continues, this problem may be alleviated.

Table 8.1 summarizes the seven case studies that follow.

CASE NO. 17

The company is a seven-year-old publicly traded company that designs, develops, manufactures, sells and services fiber-optic and other high-speed data transmission and distribution systems that connect mainframe computers to peripheral equipment, such as terminals and printers, primarily in local area

TABLE 8.1
Summary of telecommunications case studies

Firm No.	Total Assets ($ Mil)	Strategy	Reason
17	10-15	In Europe, began with distributors and slowly replaced with joint ventures for sales and service in Japan, work with trading company	Need to work closer with customer; difficulty selling to NTT; no appreciable amount of business with European PTTs; difficulty selling to Japan because of Israeli manufacturing site
18	65	In Japan, works with value-added reseller to which it provides all sensitive supporting documentation such as specifications and source code in exchange for worldwide rights (excluding Japan) to market any enhancements to the base technology developed by the Japanese reseller	Local firm helps in certification of company's product in Japan; perception of improved service capabilities in local market; commitment to sell to NTT
19	85	Originally had distributor in Europe, but terminated the arrangement after it learned that distributor had copied technology; later wholly owned sales and service subsidiaries and manufacturing facilities in Europe	European distributor copied technology and sold its own competing product; needed in-country presence to be considered a reliable supplier; service is key; "buy national" in France; pressure to manufacture locally in France; difficulty with German PTT and qualification testing; incompatible CCITT standard in Europe; difficulty with NTT standards
20	105	Uses own U.S.-based sales force to sell overseas	Difficulties due to CCITT standards used in Europe; preference for local companies shown in government procurement; need to bid through local company in Germany; perceives Japanese market as completely closed and does not pursue
21	63	Sells through an international distributor; in Japan, selected an individual to act as agent	In France, difficulty selling to PTT, which became more severe once its local distributor was nationalized; in Germany, received pressure from the Bundespost to transfer technology to several local firms; overall, nationism a problem in Europe; certain limits to the distributor relationship when times are bad; in Japan, perceived the market to be closed during the 1970s and did not attempt to enter, in the 1980s adopted a cautious approach

(Continued)

TABLE 8.1 (Cont.)

Firm No.	Total Assets ($ Mil)	Strategy	Reason
22	under 50	Uses distributors	In Germany, Bundespost has used standards to urge company to work with local company, also problem with export of sensitive data; problem of incompatibility with CCITT standards in Europe
23	under 50	Joint ventures with local partners	Uses local partners to deal with local regulatory authorities; some problems with German Bundespost which uses standards to manipulate terms of doing business; "buy national" in the United Kingdom, Germany; opposition from the KDD was overcome by local Japanese partner

networks (LANs). The company's strategy in the development of its products is to identify applications that can benefit from utilizing advantages of fiber-optic technology unavailable through conventional transmission technology. Through the use of fiber-optic technology, the company's systems provide higher-speed data transmission over longer distances with lower data error rates than is achieved with conventional cable systems. The company's systems, operating at speeds greater than 10 million bits per second and over distances of up to four miles, are oriented toward users requiring a limited number of high-speed interconnect points in a network. In comparison, the majority of existing LANs operate with large numbers of relatively low speed network interconnections.

The company has focused its efforts on the production and marketing of fiber-optic and coaxial high-speed data transmission and distribution systems for integration with computer systems currently in widespread use in large, complex information processing centers. Fiber-optic data transmission and distribution has several advantages over conventional transmission using metallic conductors. In various applications, fiber-optic technology permits increased data-handling capability over longer distances, reductions in data error rates, reduced cable congestion, enhanced communication security, and isolation from electrical interference. As a result, fiber-optic systems, in various applications, can be lighter and less expensive than systems using metallic conductor cables and can be utilized in environments and over distances where the use of metallic conductor cable is not possible.

When the company began seven years ago, U.S. venture capital markets were very tight, and the company encountered difficulty locating a source of funding in the United States. As a result, the company decided to look outside the United States for a source of financing. The company discovered that the State of Israel was offering attractive incentive packages for companies that potentially can generate high value-added exports that, in turn, help Israel with its balance-of-payments problems and shortage of hard currency. The company decided to locate in Israel and set up a wholly owned Israeli subsidiary of the U.S. parent company. For the first three to four years the company was in business, virtually all corporate functions were in Israel-- production, marketing, sales, financial, and research and development. In 1982, the company started up U.S. operations in full. Currently, 25 percent of the company's research and development and production is located in the United States.

From the outset, the company took its foreign markets very seriously. The company generates roughly 40 percent of its business offshore. Initially, the Israeli subsidiary sold to Europe directly and back into the United States only through private labels. Although the direct management of the U.S. market is now handled through the U.S. company, the European market continues to be handled through the Israeli subsidiary. The company operates this way for several reasons. First, Israel trades duty-free into the Common Market. This is not considered to be a huge advantage, but in a close head-to-head competition, that few percent has made the difference. Most importantly, the company has developed considerable expertise in its Israeli office in the area of European marketing. The company has found that the work force in Israel is more multinational in their outlook; the Israeli employees were usually multilingual and have roots in the country they are servicing. The company further believes there is an advantage to keeping the U.S. and European operations separate. There is so much good business in the United States in fiber optics right now that the European market is not often taken as seriously by most U.S. companies. Having the two groups separate allows each to put all of its energy into its own area of responsibility.

The company's initial approach to the European market was very simple. It had no base of operations in Europe, but rather dealt exclusively through European distributors. The company was pleased with these relationships. European distributors operate differently from their American counterparts; the representatives have their own service organization, inventory, and will often translate and reprint the data sheets and other product literature into the local language. The company continued with these relationships until its European business had grown large enough that it warranted more direct involvement by the company. In 1984, the company decided to approach its German distributor representative, who was doing an exceptionally good job, with the idea of forming a joint venture. In May 1984, a joint venture was formed; the company owns 75 percent and the remainder is owned by its former German representative. The joint venture is essentially a marketing and service organization.

The company then decided to form a joint venture in France similar to the German operation. The French operation, however, evolved into a different arrangement. The company has allowed a French family, with whom it had former connections, to set up a sole distributorship that is wholly owned by the family. There are clauses in the agreement stating that the company has the right to buy 40 percent of the venture at a later date.

The company is just now concluding arrangements for a wholly owned venture in the United Kingdom that will also be a marketing and service organization. In addition, the company has a small support group that it is also moving to the United Kingdom. This is a group of sales and service experts whose job is to support the other European representatives. The company's other European markets, Denmark, Sweden, Belgium, and Holland, will continue to be serviced through the existing distributor relationships with needed support coming from the U.K. support group and Israel.

The company has been very successful in Europe. Roughly 30 percent of the company's worldwide sales are generated in Europe. Its customer base in Europe is principally manufacturing companies, banks, and universities. The company has not done any appreciable amount of business with the European PTTs. At this stage, the company perceives that it is impossible to break into these markets unless a firm wants to spend large amounts of money pursuing that market and even then, results can be disappointing. Fortunately, the LAN market has been and continues to be open in Europe. As a result, the company is able to do good business in Europe despite the closed nature of the local PTTs.

The company did not approach the Japanese market until 1983. For the first three to four years the company was in business, all of its operations were based in Israel. This made it essentially impossible to sell into Japan; the Japanese are very sensitive about doing business with Israeli-content companies because of their tremendous dependence on Arab sources of oil. The company also believed that in the initial years of its growth, it had not yet developed sufficient infrastructure to support a Japanese market. After 1982, when the company set up full U.S. operations and became a public company, it began to get inquiries from Japanese companies.

In 1983, the company decided to allow a small, but resourceful, U.S. marketing company to make inquiries on its behalf in the Japanese market. The firm was employed on a retainer basis and was given exclusive rights to market in Japan for about six months. During that time period, the marketing firm did manage to bring a few potential Japanese customers to the company, but the company was unable to make a sale. The company then decided it would have to handle its entry to the Japanese market itself. It attributed these initial difficulties to the monopolistic hold that NTT had over the Japanese telecom environment. Nonetheless, it was confident that its products were unique and believed that the Japanese market would have to eventually open up.

In 1984, the company began directly contacting Japanese firms that it had targeted as potential sellers of the company's equipment. In the early months of 1985, a Japanese distributor began to show genuine interest. This Japanese firm believed that with the privitization of NTT and the opening up of the market, it could be in a good position to take significant business away from NTT; this business previously was unavailable to private industry in Japan. After much negotiation, the two companies recently agreed to a three-month trial. This Japanese distributor was given exclusive rights to market selected products for three months in Japan using the company's name on all advertising. In this way, if the Japanese firm decides not to continue with the company's product line, at least the company would gain free exposure, and thus, would not be losing valuable time.

The company believes that a relationship with the right distributor in Japan will provide the company with an adequate presence in the market. Basically, the bottom line is profitability; if the company can penetrate the market by shipping products FOB to San Francisco or Tokyo directly and running a symposium there twice a year for its salespeople, it can save itself the enormous expenditures associated with setting up operations in Japan. If the market were to become very large there, the company might make a greater financial commitment to the market. It is estimated that Japan will be a $3-5 million a year market for the company in a few years. If the products are right for Japan, and the distributor is right for the company, the market could be much larger.

CASE NO. 18

The company, founded in 1968, is a leader in data communications, manufacturing equipment, and designing networks that enable information to be moved and managed more efficiently. The company is a pioneer in the field of data communications, specifically in the technology of statistical and time-division multiplexing, switching, network management, and control. The company supplies sophisticated data communications equipment for large, complex networks with hardware installed in many of the 500 largest U.S. corporations, including AT&T, Citibank, RCA Global, Sun Information Systems and Digital Equipment Corporation. The company's overall strategy for future growth is to successfully evolve into a fully market-driven firm with an emphasis on providing a completely integrated communications network for the customer which combines elements from the company's entire product line including modems, statistical multiplexers, high-speed time-division multiplexers, intelligent switching systems, and network concentrator and processors. Manufacturing facilities are located in the United States and Puerto Rico. This year, the company will generate roughly $90 million in sales, with sales outside the United States at $15 million, and $2 million in sales in Japan.

The company has been involved in the Japanese market since 1980, when it established a distributor relationship there with a value-added reseller. The company went with a distributor relationship because in 1979 the company had just cleared $10 million in sales and therefore could not afford to establish a wholly owned subsidiary in Japan. In addition, the company did not want to have to involve itself in the day-to-day operations of a venture in Japan; the company wanted to leave those matters to a Japanese partner. Further, the company believed the size of the potential market did not justify a more direct and involved foray into the Japanese market. The Japanese distributor handles the company's entire product line; although it acts as an agent for other companies, it carries no competing lines.

From the beginning, the company decided it would establish a completely open relationship with respect to its technology. The company and its Japanese distributor have a contractual agreement that gives the Japanese company access to source codes and software listings for the company's products, information that is very closely guarded from the company's competitors. This gives the Japanese value-added distributor the ability to modify and enhance the company's products, in order to respond to customers' and market needs quickly and efficiently. In exchange for this technological base, the company was guaranteed the worldwide rights to market any enhancements to the base, and the Japanese distributor was given the right to market the base product and any enhancements in Japan. The agreement includes provisions that require the Japanese partner to provide the company with any modifications or enhancements to a product it develops. The company trusts that its Japanese partner will protect its source code and it is not concerned that the Japanese partner will develop into a competitor; the present arrangement is quite lucrative for the Japanese partner.

The company has been very satisfied with its level of sales in Japan thus far. No problems have arisen with respect to the certification of its products in Japan. The company's distributor has applied for exemptions for the product on a case-by-case basis and approval is received within a few days. The company believes that its decision to form a relationship with a Japanese value-added reseller and to provide that company with sensitive source codes and software listings was instrumental in gaining the NTT contracts. Having access to source codes and software listings for the products allows the Japanese distributor to offer a much better level of service by making on-the-spot complex problem solving possible through the Japanese distributor. In the area of data communications, this service capability is critical. Since the company's competition is largely other U.S. firms that deal through distributors but do not provide sensitive source code data and software listings, there is a perception in the Japanese market that the company is able to offer better service. Consequently, the company is in a good competitive position in the market.

NEC is the company's main Japanese competitor, but in several areas NEC has not yet developed the capacity to offer products technologically on a par with the company's. In cases where the customer does not have the time to wait for NEC's "paper product," the company is faring well. NEC does have one product that directly competes with the company, and in that product market NEC already has a 27 percent share. In this case, NEC was able to bring the product to market six months ahead of the company and quickly took the market lead. Overall, the company is optimistic about future prospects for the Japanese market because of the increased interest in data transmission and the boom in the computer hardware market: the company's Japanese sales could grow to $5-$6 million over the next two years. Of this growth, probably one-third would be to NTT and the remainder would be to private companies and banks.

The arrangement between the two companies is now five years old. During this time, any and all enhancements made by the Japanese distributor to the base technology provided by the company have been transferred to the company. This has worked very well for the company; in several cases the Japanese partner's developments have significantly outpaced parallel developments carried on simultaneously by the company, and the company has received this technology at no cost. The company is very happy with its current arrangement and foresees no changes in its strategy toward the Japanese market within the near future. If the Japanese market were to become a large and significant market for the company in the future, the company would consider moving manufacturing to a site in the Far East, but it probably would produce in Taiwan or Hong Kong to be most cost effective.

CASE NO. 19

The company is a twenty-six-year-old publicly held firm that manufactures and sells electronics-based test and measurement systems for the areas of data, microwave, and video communications. The company is highly decentralized, with a small corporate headquarters and over thirty sales and manufacturing companies throughout the United States and abroad. Last year, the company generated over $200 million in revenues.

The market served by the company's data communication group is composed of large companies that have an increasing need not only to computerize operations such as payroll and manufacturing, but also to have their computers and peripheral equipment communicate with one another from remote locations. Industries such as airlines and banking with worldwide branches cannot exist without complex data communications centers. The company offers a comprehensive systems approach to the market, involving varied data communications product lines. Nearly one-third of this group's revenues was from overseas sales in 1983; this figure dropped to roughly a quarter in 1984 due to the strengthening of the U.S. dollar abroad. The company believes it is essential to establish a country-by-country presence to market effectively in international high-technology markets. This past year, the group generated roughly $100 million in sales worldwide.

The company began selling to Europe in the early 1970s through a Swiss distributor that sold the company's products in several European countries. The company later discovered, however, that the distributor sold the company's product directly for a few years and then designed its own product using the company's product as a model, had it made, and began selling its own product to customers. The company was unaware this was happening until it received a purchase order direct from a

European customer and filled the order. The customer then wrote back asking why the equipment was different from the equipment it had received from the company's distributor. The distributor's "new" business was thus uncovered, and the relationship between the two was quickly severed. The company did not take the distributor to court because, given the amount of business the distributor had taken away, it was not worthwhile. The distributor continues to compete with the company, but is is only generating a few hundred thousand dollars in sales per year.

In the middle 1970s, the company started its own wholly owned sales, marketing, and manufacturing facility in Europe. The main motivation for this was the need to offer the market something that the company's U.S. competitors did not. None of these firms had any form of direct in-country presence. The company decided that it would offer service. The company believed that customers were frightened off by companies that did not appear to have a commitment to the European market. Over the years, customers had expressed concern over service and continuity. The company believed it could overcome these impediments by establishing its own marketing and manufacturing facilities in Europe. The manufacturing facility was established and produces all the same products manufactured by the company in the United States.

The local PTT authorities in Europe must certify the company's products. The company encountered difficulty certifying its products in Britain. The company originally manufactured outside the United Kingdom and it was not until the company acquired an English firm that it was able to qualify its products.

Since the company began manufacturing in Europe, it has noticed a much greater acceptance of its products, but still encounters a strong preference for local products. In the company's test equipment product line, there is a French firm manufacturing similar equipment. That French firm wins about 90 percent of the contracts, even though the company's product line does very well in other countries. As a result, the company has begun to do some assembly work in France to build some French content into its product. The company perceives that this problem will persist until it sets up a full manufacturing facility in France.

West Germany has presented problems of its own. The German PTT controls the qualification of a wider scope of telecommunications products than the other PTTs in Europe. While the control of other postal authorities stops with line equipment, the German PTT's authority continues up to and including terminal equipment. Furthermore, the German PTT subcontracts its qualifications work to private companies such as

Siemens, Philips, and ITT. Needless to say, it has not been easy for the company to gain qualification with its competition performing the qualification tests. Besides the obvious preference for locally made equipment, the company also has encountered problems with CCITT specifications. Because CCITT standards differ from American standards, the company has had to technically educate the inspectors in order to be qualified. The company estimates that time lost in the qualification process has cost the company roughly $500,000 in lost contracts and opportunities in Germany over the last few years. Moreover, when it became known that the German PTT refused qualification of the company's products, the effects of the resulting loss of credibility were felt beyond the German marketplace. The company found that it had to explain these certification difficulties to other customers as well.

Despite these problems, the company has done well in Europe. The company's European sales have grown from $1 million in 1977 to nearly $20 million this year. When the company started, it had roughly 5 percent of the European market; today it has between 30 and 40 percent. These gains have been made even though U.S. companies as a whole have dropped from 80-90 percent of the market in 1977 to about 55 percent today. The company has done well against its U.S. competitors. It believes that its success is a result of having local manufacturing sites. None of its U.S. competitors has set up its own manufacturing facilities in Europe.

The company employs no Americans at its European subsidiaries. It believes that local support manufacturing, coupled with an understanding of the local culture, is essential to the success of the organization. It also has found that workers tend to cooperate better with their own people than with Americans.

The company believes there are several reasons for its overall success in Europe. First, the business was relatively immature when the company originated, and subsequently it encountered little competition/opposition. The company also is involved in a highly specialized, low-volume market niche. This market segment is not very attractive to the larger companies. Further, the European PTTs are more concerned about and tend to focus their attention on the transmission side of the business; the company's products address the digital side of the business. Moreover, the company has been able to offer a product that is technologically advanced, and equivalents are not readily available from local sources. The company believes this last factor to be most important. "Buy national" policies run very strong throughout Europe in the telecommunications industry, but these prejudices can be overcome if a nonlocal company has a product that is technologically advanced, not locally available, and for which there exists a need in the marketplace.

The company has not met with equal success in Japan. It has been trying to sell into Japan for the same amount of time and is only doing a small fraction of the volume it generates in Europe. Although the company is generating nearly $20 million in sales in Europe this year, it will generate, at best, $1 million in Japan this year with the same product line.

The main impediment has been NTT. In particular, the company found that NTT favored local companies over non-Japanese companies through the establishment of specifications. One particular instance was given in which the company was trying to sell test equipment. There were several companies competing; all were American. NTT then issued a set of specifications that was developed--had some of the specifications of each of the competing products, with the result that none of the American firms complied, including the company. Soon after, a Japanese company introduced a product that exactly complied with the specification, and it made the sale.

The company expressed a great deal of concern over having its product copied on the Japanese market. The nature of its product is such that it is, with a fair degree of effort and expertise, easily reverse-engineered. It also is not patentable because it is a standard set of components coupled with sophisticated software. As a result, the company has not wanted to work through a distributor or joint-venture agreement for fear of losing its technology.

With the recently announced split-up of NTT and alleged opening up of the Japanese telecom market, the company sees new opportunities in the market. The company estimates the potential Japanese market for its products to be close to $20 million. The company is now considering establishing a wholly owned manufacturing, sales, and service subsidiary in Japan. The company believes that if it manufactures in Japan and uses Japanese nationals to do sales and marketing, it has a much better chance to excel in the market than if it tried to supply the market from the United States. The company soon will decide whether it is willing to make the $2-$3 million investment.

CASE NO. 20

The firm, a publicly traded company, is actively involved in several areas of telecommunications industries: currently, it manufactures three types of digital and analog transmission equipment including satellite earth stations, troposcatter and line-of-sight equipment, as well as unique and standard digital multiplexers, digital speech interpolation equipment, time division multiple access, and supervisory control systems for telecommunications networks. The company's equipment systems are used to transmit and receive telephone, television and telex signals, computer data and other monitoring information. Within the computer field, the firm manufactures color monitors, terminals and computers; and provides computer systems and software for applications such as process control, military command and control, telecommunications networks, and computer-aided design. Other equipment manufactured includes radars, radar subassemblies, avionics and airborne telemetry equipment, and support components. The company employs approximately 2,000 individuals.

For satellite communications, the firm is a major producer of analog and digital equipment including high-power amplifiers, low-noise gallium arsenide FET amplifiers, up-and-down converters, and satellite monitors. It has developed state-of-the-art equipment to enable more efficient use of a single satellite transponder called time division multiple access (TDMA). TDMA is a key technology in the field of satellite communications networks. In TDMA, transmissions or earth stations accessing a particular satellite share the transponder in time by transmitting their data traffic in bursts. By employing proper synchronization methods, transmissions are sequenced for cross-communications without overlap or interference. TDMA is especially useful in transmitting high rates of digital data or voice.

In addition to manufacturing transmission hardware, the company provides turnkey satellite earth stations for international and domestic voice, video, and data transmission applications for common carriers and private networks. These stations, both digital and analog, are implemented on a turnkey basis from conceptual design and hardware manufacturing through field installation and commissioning.

The company's satellite communications group maintains an aggressive international sales force. At its own admission, however, this group is involved in no satellite earth station business in Japan. The company perceives that the Japanese market is closed to them, although no specific events were cited to support that view. A decision was made some time ago to make no efforts in Japan because the company believes that Japanese equipment is technically on a par with the company's equipment. Furthermore, the company has been unable to obtain sufficient information from the procuring bodies about the required equipment to produce a competitive bid.

In terms of standard satellite technology, American companies have no advantage here. In fact, the Japanese have already begun to speak about the development of telecommunications for the KA band, which is a much higher frequency than that used in the United States. American companies are unwilling to sink R&D funds into programs such as these because of the restricted market and less profit potential in the near future.

The company also has experienced difficulties obtaining sufficient detailed information from the procuring bodies in Japan. To a large extent, a solid competitive bid in the satellite communications industry is the result of detailed understanding of the systems and equipment requirements, and previous collaboration between the manufacturer and the procuring body is very important. In this area, the Japanese manufacturers enjoy a substantial advantage. The reluctance on the part of the Japanese to consider foreign suppliers when local industry produces all the required equipment makes competitive bidding uneconomical and unattractive for U.S. suppliers. Based on these considerations, the company decided not to compete in Japan for satellite earth station products.

The company's involvement in Europe also is minimal although it perceives that the European market will be expanding. Several European countries are beginning to establish their own television distribution by satellite, and several new satellites have been launched in Europe. Nonetheless, in Europe the total business is divided by country, and each country (especially France, Germany, and to some extent, Italy) is determined to have local requirements met by its own fledgling telecommunications industries as much as possible. Hence, it has been difficult for the firm to break into the European satellite communications market for several reasons: in part, it has encountered discriminatory procurement of local equipment, examples of which are described below. Moreover, in many European countries (Germany, Belgium, etc.), the company has had to submit its bids through a local company.

There also is a problem of nonconforming standards. European telecommunications equipment is manufactured according to CCITT standards. Although the U.S. standards for the same equipment are considered to be more stringent, both sets of standards are sufficiently different to result in an inability in many instances to achieve compatibility between both sets of equipment without considerable modification. Intelsat, for example, requires CCITT standards for the interface between the user and the earth station in the TDMA systems they are introducing. In some cases, the company has bought European-made CCITT conforming equipment and retrofitted it to its own system to circumvent the interface problem. In the company's opinion, the costs of modifying the equipment, such as TDMA, have prohibited it from bidding on contracts unless there was a very good probability of winning the contract and a strong prospect of long-term follow-on work on the system.

The insistence on CCITT standards sometimes creates difficulties in selling American-made equipment not only in Europe but also in certain areas of ex-colonial Africa and other developing nations where a preference for CCITT equipment is strong. In these areas, the lack of indigenous engineering capabilities also works against the company because the procuring bodies have insufficient technical understanding of the issues involved, which in turn makes it difficult to present alternative solutions.

In December 1983, the company learned that a newly established British common carrier wanted to acquire a satellite earth station to communicate directly with the United States. The contract for this system would have been worth about $1 million. The company was involved in this project from the beginning; several visits were made to England to work with the customer to ensure that the proposed equipment satisfied the requirements. After the bids of the satellite earth station were submitted, the customer contacted the firm and stated that additional equipment would be required and asked for a price for this additional equipment. The firm replied that the additional equipment would add only $1 to the bid price. The customer was pleased and the company felt very confident it would win the contract. After three or four weeks had passed, however, there was no word from the customer. At this time, the company contacted the customer in England and learned that the contract had been awarded to a British company. The company then contacted its own sources in England to find out why it had lost the contract. Information received pointed to the possibility that during the evaluation process import duties, payable directly by the customer, were added to the company's bid price, resulting in a final price that was slightly higher than that of the British firm. Although the company could not substantiate the claim, it felt at the time that it was being discriminated against because it was not a British company.

In late 1984, the German government issued an RFP for a large number of small earth stations with a value of approximately $1.25 million. The RFP stipulated that all non-German companies had to bid through a local German company. Therefore, the firm submitted its proposal to a company in Bonn, which, in turn, submitted the company's proposal as part of its own proposal to the government. Six weeks later, the firm received word that the procurement was most likely going to be divided in three because the government believed the order was too large for one contractor to produce and deliver on time. At that time, there were only three companies bidding on the job--the company and two German companies. The company did not adjust its price per unit to reflect the reduced volume; it simply divided its original bid price by three. Several weeks later, the company was informed that the entire contract was awarded to a German firm. The company then contacted the firm in Bonn through which it had been bidding and asked for an explanation of what had occurred. This German firm claimed it had asked the government for a seven-day extension in delivery of the proposal and was assured that this was not a problem. Then, after submitting the proposal within the seven-day extension period, it was told that the proposal was late and would not be accepted. Assuming that this explanation was indeed true, it could only be interpreted as another case of preferential government procurement practices.

Similar problems have been encountered in Canada. In a recent procurement, the firm was told that a Canadian company might receive preferential treatment over a non-Canadian company. The firm then asked the Canadian company with which it was bidding if it would help its position if it bid with a Canadian company as the prime contractor. The company was told that it might, but the government may require that the equipment be manufactured locally even though the price of the Canadian supplier may be higher.

The incidents cited above indicate the difficulties encountered in bidding to foreign countries where strong preference for local contractors exists. This must be accepted as an understandable, although from the company's point of view undesirable, attitude on the part of certain foreign countries and factored into marketing plans.

The difficulties encountered in the European earth station market resulted in a change in foreign marketing strategy of the company, with more emphasis on Third World telecommunications markets. Company sources have stated they do not expect the company to make any entries into the Japanese market within the foreseeable future. The company's chief interest remains in the U.S. market. If the company spends any money on research and development, that research will be oriented toward the U.S. market. The company does not intend to develop equipment specially for the European market; it will continue to sell overseas only equipment that can be adapted at low cost and minimal effort.

166

CASE NO. 21

The firm is a publicly held company involved in the field of telecommunications. The company designs and manufactures computer-based systems that address telephone operating company needs in the basic areas of billing, switch enhancement, traffic/network systems, operating services, and subscriber service. In the area of billing systems, the company provides a high-capacity centralized data collection system designed for both electronic and electromechanical switches that help telephone companies deal with the increased complexity and volume of billing data generated by local measured service requirements, as well as a system that addresses the billing needs of older step-by-step switches, and a system that provides local and long-distance detailed billing through an upgrading of crossbar switches. The company's traffic/network systems furnish telephone operating systems with traffic engineering data and network performance data. Like the billing and traffic/network systems the firm offers, most of the company's products are ancillary to the telephone network and therefore are not necessary for providing basic telephone service. Rather, these products enhance the level of service a telephone network can offer the customer.

At present, the company has products installed in roughly thirty-six countries around the world, not including the United States and Canada. Its customers include most of the telephone administrations in Europe, the East, Far East, South Africa, and many in Latin America. The company first decided to market its products abroad in 1973. The strategy the company elected at that time was to ally itself with a major international telecommunications firm that could provide the company with worldwide representation. The company entered into a relationship with International Telephone and Telegraph (ITT), which was very strong in Europe, Latin America and the Far East. ITT became the company's worldwide representative with the exception of the United States, Canada, and Japan, where the company decided to handle the marketing and sales effort itself. Overall, the company has been pleased with this relationship.

According to sources at the company, the decision to go with ITT was shaped by several considerations. With only $12 million in sales in 1974, the company believed it did not have the financial resources to involve itself more directly in these markets either in the form of wholly owned subsidiaries or joint ventures with local companies. The primary reason for this "alliance," however, was the fact that the telephone industry is "protected" in most countries, and the quickest way to do business, in fact perhaps the only way, is via a known, accepted supplier to the telephone administration in each country. ITT was already well established in Latin America, Europe, and the Far East. The company believed it could get the maximum exposure for the least financial outlay through a liaison with ITT. As a result of this liaison, the company did very well in Latin America during the 1970s, and has done exceptionally well in Spain and Portugal, and lately in Sweden and Holland. The company attributes its success in these areas to two factors. First, at the time it took its products abroad there was no indigenous competition in these areas; and second, the products they offered were ancillary products that were not basic to providing telephone service, and as such were not considered to be a strategic resource. The company believes it has fared rather well in countries where its products were not competing with ITT's. In Germany, France, Italy, and Sweden, where ITT had manufacturing facilities, the company did not do as well. The key to its business was based on its ability to "sell" the individual ITT companies, on a country-by-country basis, that the sale of companies' products would not negatively affect its (ITT's) business and would, in fact, enhance ITT's reputation with its customer.

The company has never been able to make any real progress in France. In fact, the first time the company went to a French telephone exchange with equipment to give a demonstration, the company's representatives were literally thrown out of the office; this exchange supervisor did not want to buy American equipment. This was the mid-1970s. In France, the company tried to sell its products to the telephone administration through LMT and later through CGCT, ITT's French subsidiaries during the period under discussion. During this time the company made no significant entries to the French market; it has only one system operating in France. The company believes that there was, and may still be, a definite resistance on the part of the French telephone industry to buy American. Since CGCT was taken over by the government a few years ago, the company has made no further attempts to penetrate the French market.

In Germany, the company sold through Standard Electric Lorenz (SEL), ITT's subsidiary there. Germany, like France, has certain accepted suppliers to the telephone industry and a strong preference is shown for locally made products. The market there is divided among SEL, Siemens, and TNN. The company perceived that SEL would actively market its product only in cases where SEL was in need of revenue, did not have a product of its own to sell, or saw some other political or business benefit. If the company's product had to be sold at the expense of SEL's own business or market share, then SEL did not try to sell the company's product.

Within the last five years, the German Bundespost announced a competitive tender worth approximately $30 million for a selective line billing system, equipment that SEL could not supply itself. SEL was interested in pursuing the contract because it had recently lost some business and was in need of additional revenue. As a result, the company was able to place a bid for the selective line billing equipment through SEL and won the contract in competition with Siemens and TNN.

Prior to submission of the tender, the company was approached by SEL with a request for a written agreement that would provide for the granting of manufacturing rights to SEL, Siemens, and TNN. SEL claimed that the German Bundespost would allow no equipment to be used in its network unless local suppliers had the right to manufacture the equipment; SEL insisted that the option to manufacture would most likely not be exercised by any of the German firms; it was simply a precondition of the contract. An agreement was reached wherein the first X percent of the project would be guaranteed to be manufactured by the company; during the second X percent, there could be a gradual transfer of manufacturing know-how to one, or more, of the three German firms; and for the remainder of the project, the German firms would continue on their own with royalties flowing back to the company for software, etc. Another requirement of the contract, however, was that the company's system had to connect to a Siemens computer. After a year-and-a-half of trial with constant delays due to the lack of proper interface with the Siemens computer, together with the rising dollar, the customer lost interest in the project and withdrew funding. As a result, the issue of technology transfer was never resolved. The company attributes this failure to two factors: Siemens' apparent unwillingness to cooperate with the company, and SEL's unwillingness to apply pressure to help settle matters. SEL would not risk alienating either Siemens or the Bundespost engineers, who also seemed to be operating in a foot-dragging mode.

The company's experiences in Sweden, Germany, France, and Italy have all been quite similar. These countries have strong indigenous telephone equipment manufacturers. As a result, the company has found it very difficult to contract within those countries. In the cases where the company has done business in those countries, the company believes it was due to a stroke of luck or a local representative badly in need of revenue. Norway was such a case. When the sales of the local ITT subsidiary, Standard Telephone Kablefabrick (STK), began to drop, the firm needed to look elsewhere for sources of revenue. STK then began to actively push the company's product in Norway and was quite successful. However, when STK's product line was in better shape and sales were back to previous levels, its marketing efforts for the company's products diminished, and sales dropped off. Again, this typifies much of the company's experience in these countries. Similar tales could be told of Holland and Sweden.

The company has only once considered manufacturing in Europe; this was in conjunction with a proposed detailed billing project in England. If the company won the bid, it intended to set up a manufacturing facility in England that would supply the United Kingdom and Europe. The company did not believe that this manufacturing facility, which would provide a source of production internal to the common market, would make it easier to sell into other EC members such as France or Germany; each country within the market is very protective of its indigenous telecommunications industry, and nationalistic preferences run high within each country. It did believe, however, that the U.K. manufacturing facility would give it a source of product denominated in the pound sterling, which would help the company to price its products more competitively within Europe, given the dollar's recent strength. The British government also offers attractive export financing to companies trying to sell products abroad, and the company believed that this could aid the company in Third World markets, where the United States does not offer comparably attractive financing packages.

Unfortunately, the company did not win the contract. The company, which had already installed a fairly large billing system in the Community of Bristol, was asked by British Telecom to submit its bid directly rather than through the local ITT subsidiary. The project, which would convert all of the existing equipment to provide detailed billing for all subscribers, was to be in the $300 million range and was to be split between three companies. It was obvious that two of the companies would be British; however, the company was led to believe it was British Telecom's choice for the third and, therefore, was already in the process of locating an in-country manufacturing site. Suddenly, the company was notified that the last third of the project was awarded to a very large American multinational corporation that had seen this as a very large market, one with no demonstrable experience in this field, but one with considerable political clout. After this, the company abandoned the idea of a European manufacturing site.

The company approach to Japan has been different. The company made no attempt to enter the Japanese market during the 1970s because it perceived that the market was closed and did not want to waste time, energy, or money. The company's marketers began to make trips to Japan in 1979, initially with the intention of making contacts and finding an agent or representative for the firm's products in Japan. The company began by approaching some of the large trading companies like Mitsui, but these discussions proved unsuccessful. Eventually, the company made contact with a U.S.-educated Taiwanese who was very well connected in Japan, and it decided to establish him as the company's Japanese agent in 1981. For the four years that followed, the company sent representatives to Japan several times a year for audiences at NTT. After much hard work, persistence, and patience, the company was invited by NTT to submit a proposal for billing meter observation equipment. The company was told by NTT that although NTT would not buy large quantities of the meter observation equipment, if the company did a good job and there was a subsequent change to detailed billing, then the company would be in a good position to win the contract for detailed billing. The proposal was accepted by the NTT procuring body, but the NTT engineering group insisted on a continual round of technical discussions, which led to a change in the technical proposal in order to gain their concurrence. Unfortunately, by this juncture, the company began to experience financial problems, due to other business factors, that made it difficult for the company to make the needed changes because of the R&D funds involved. Currently, the company is still negotiating with NTT, and it appears that NTT is reluctantly accepting a company-proposed alternative.

CASE NO. 22

The company is a five-year-old privately held firm that provides communications-related proprietary software and services. The company's software is directed toward the following market segments: consulting, facilities management, contract technical services, brokerage vertical market, financial services, and the insurance industry. The company generates the majority of its revenues from domestic sales; less than 10 percent of total sales come from overseas.

The company first did business overseas in 1982. It provided and installed services in Italy that were contracted in the United States. To date, this is the only tangible business the company has contracted overseas. Later that year, the firm faced substantial difficulty in establishing itself in Germany. This was due to two problems related to standards. First, the German Bundespost would not give the company frequency clearance for a "downlink" into Germany, which would enable it to provide data services in Germany that were transmitted from a computer in the United States via a direct broadcast satellite to a packet-switched network. Second, the German government would not approve the data base the U.S. firm wanted to provide because of concerns about the export of sensitive data about the German economy.

The company felt that the German government had ulterior motives. It believed that the German government was using standards to try to get the company to team up with a German firm and then to have the company teach the German company what it was doing. However, the German government wanted the company to put up all the capital and did not want to buy any of the R&D that the firm had already done to compensate it for its expenses.

The company spoke to several German companies about such an arrangement and approached them about establishing a joint venture. However, when it drew up the memorandum of understanding (MOU) necessary to incorporate, the German firms were unwilling to provide commitments of capital. The company felt that even with a successful joint venture in place, it would have had no guarantee of access to the market.

The company then approached the German Bundespost. An executive spoke with the head of the telecommunications policy section, recounting the barriers the company faced. The representative of the Bundespost spoke with the potential joint-venture partners the company had seen. Executives from these German firms then became resentful that the American executive had approached the government.

Standards and government requirements became key determinants of the firm's overall performance because the company had to maximize its profits by expanding the use of the services it designed, in order to get the R&D dollars invested in the product returned. Because standards such as those imposed by Germany forbad the company from deploying a data base service without substantial modifications, and made it undertake major efforts to find a local partner, the company's cost of business became too high, and it could not operate in Germany.

The company spent a considerable amount of time and funds before it sought to enter the German market, developing a way to get around European (CCITT) standards that are incompatible with those established by the United States for satellite transmission. The United States uses 56 kilobit bandwidths, while the Europeans use 50 kilobits. Thus, there is a built-in incompatibility between U.S. and European systems at the physical interface level. Therefore, the company had to design new protocols to overcome this incompatibility.

Although it succeeded on a technical level, it has been unable, thus far, to convert its technical innovation into profits because of the pressure from the German government. In addition, the company's executives felt that the two years it had spent in developing the access technologies that permitted it to offer such a data base service in Europe has been wasted. The company decided to refocus its efforts on the United States and Canada.

With the recent privitization of British Telecom, the company is now considering forming a joint venture with a British distributor. Although the company has not had any success in Britain in the past, it believes that there could be a sizeable market for its products and services now that British Telecom has changed hands. The company is currently considering this matter; a final decision has not been made.

CASE NO. 23

The company is a four-year-old privately held firm operating in the telecommunications industry. The company conducts two types of business: managing telecommunications projects, and developing projects and companies in the telecommunications field in return for an equity stake. Roughly 65-75 percent of the company's business is generated outside the United States. Since its inception, the company has been involved in fifteen development projects worldwide. The company's main areas of business outside the United States are Asia (excluding Japan) and Western Europe (United Kingdom, Italy, Belgium, the Netherlands, and Luxemburg).

From its beginning, the company has never made any distinction between the domestic and overseas markets in terms of corporate priorities. The company was involved in both markets from the start. The company believes that the largest barrier to doing business abroad effectively is in obtaining sufficient information about a project and understanding the information that is available. To ease these difficulties, the company bids on projects as part of a team that usually includes a local firm that provides a local point of view and understanding of indigenous business practices and regulatory bodies. Generally, the company prefers to work with a manufacturing company that can provide most of the technology. All the company's overseas work is performed as a partnership activity. Outside the industrialized world, countries are interested in whole system, turnkey projects. The company believes that as a small business it can do best in these markets by being part of a project team; the company has to sell itself only to the other members of the team. The project, as a whole, markets itself.

The company has experienced a fair amount of success in Italy, the United Kingdom, and the Benelux countries. Doing business in these countries, however, has not been without problems. The Benelux market is relatively open. Whether the company is selling to the Dutch PTT, Philips, or the Belgium PTT, it has found that if a mutual advantage for both parties can be demonstrated, then a reasonable deal can be put together. In the United Kingdom, the company perceives that there is still some pressure to "buy British" on the governmental level. This is particularly evident in any project that involves long-term procurement of services or facilities operation. Overall, procurement is open because the United Kingdom recognizes the importance of a top-grade telecommunications system and network to the country's well-being. In Italy, the company has found it difficult to negotiate because it is never clear which local entity is in charge. The market, however, does tend to be competitive. The company has found that as a service (design and management) provider it does well in Italy because it can easily provide local content in a project.

The company has not had any substantial success in the German market where the German Bundespost has made it almost impossible for the company to do business. The company has found that, in Germany, the desire to protect certain telecom markets does not necessarily bear any relationship to what is economically important to either the German firms that participate in these projects or the German economy as a whole. There is a certain irrationality involved in these markets that makes it very difficult for companies to do business. The company has encountered considerable difficulty negotiating operating standards with the Bundespost; in several cases, the Bundespost tried to levy requirements on a proposed system, which were clearly inappropriate and unnecessary. At each step of the way, the Bundespost has requested either implicitly or explicitly that the company offer something more--either equity, a greater share of the profits from a particular service or application, or a certain amount of work for free (most common). These difficulties exist throughout the Bundespost, at all levels of the bureaucracy. On a few occasions, the company had approached more senior members of the organization for help on a particular project. Although the senior members assured the company that the matter would be cleared up, the company continued to encounter difficulties. Since the Bundespost is a tight organization in which senior management is able to effectuate quick change, the company has concluded that these groups are in collusion.

The company worked on one project that involved establishing a European network for a major consortium. Though initially only one of the two large operators was to be in Germany as a result of pressures from the German government, two German operators were eventually included in the project. Nonetheless, the project still did not receive approval from the German Bundespost. This particular project would have provided for 1,100 local jobs in Germany and several million dollars in domestic investment.

Experiences such as these have led the company to conclude that the pursuit of the German market is not economically feasible or worthwhile. The company does believe that this may change in the future. Despite its excellence in other areas of technology, Germany does not have a dominant provider in the telecommunications industry. Siemens is not competitive on a purely technological basis. Increasingly, these weaknesses are beginning to be associated with Bundespost policy. For example, the Bundespost decision not to adopt a digital system in the mid-1970s meant that Siemens did not move resources into the development of digital technology. Today, Siemens' digital technology is a generation behind its competitors' equipment.

The company has found the Asian market an easier one in which to do business. The company is presently involved in a fifty-fifty joint venture in Southeast Asia. Originally, the company was allied with a major North American manufacturer on this project. Approximately 60 percent of the equipment used was to be North American technology. The design of the project, however, necessitated a certain amount of modification to the manufacturer's standard equipment line. The manufacturer would not agree to make these modifications because it was unwilling to cut its profit margins and had become discouraged by time delays. As a result, the company dropped this alliance and only 20 percent of the technology employed on the project will be from North America.

The company's present partner is a domestic firm. The firm provides the project team with an understanding of the limitations of the local environment including the physical, technological, and political aspects within the country. The company uses this information to help choose appropriate technology and effectively manage and execute the project. This project does entail a fair amount of technology transfer. However, the company maintains that though the transferred systems embodied leading-edge technology when the proposal was written, these systems will be outdated by the time the project is finished.

The company attributes its success in the Southeast Asian market to its forthright approach and commitment to building strong local relationships and partnerships. In addition, the company has sought to understand the local needs and to view those needs as a pragmatic basis for designing the project. Whereas other companies want to sell their product as is, or sell their own design approach, the company has always exhibited a genuine willingness to modify its designs to meet customer specifications. As a result of the company's flexible approach, the government has extended the original project. Rather than performing a four-year, $60 million pilot program, the company and its partner received a contract for a three-year, $300 million project covering five urban areas.

The company has had varied experiences with regard to technology transfer. In one instance, the company was involved in the sale of terminal equipment to Korea. When the contract for the sale was being written, the company was aware that certain sections of the document were sufficiently vague that the matter of supporting documentation was left unclear. Nonetheless, the company (and the manufacturer it was working with) opted to proceed with the sale as planned. Not surprisingly after the equipment had been manufactured and delivery had begun, the Korean customer pointed to the vaguely written section in the purchase agreement and requested certain design specifications and drawings. The Korean customer made it clear that if the drawings and specifications were not delivered, that shipments were to cease. The company did not want to involve itself in a lengthy legal suit and decided to make a deal with the customer for only a partial transfer of technology.

The company does not believe that technology transfer is necessarily bad. The company worked on one project that involved a major U.S. computer hardware and service provider. The company strongly recommended that this firm join forces with a Japanese electronics firm; some exchange of technology would be necessary. The U.S. firm adamantly refused to involve itself in any type of technology transfer and backed out of the project. The company viewed this as imprudent because leading-edge technology was not involved and, in addition, innovation keeps a firm ahead of its competition.

The company also has another large project in Asia in which the major partner is a Japanese electronics manufacturer. The company's relationship with its Japanese partner involves only this one project. The Japanese firm actively sought out the company's expertise for this project. The Japanese firm had actually competed against the company for the Southeast Asian project and failed. Since the company had succeeded where the Japanese firm had failed, the Japanese firm approached the company about some general business opportunities. One year later, the Japanese firm asked the company to work with it on this particular project. The Japanese company was attracted by the company's flexible approach to systems design; the Japanese approach to technology is fairly rigid and this has lost the Japanese firm some business in the past.

The project, which is for the development of an international communications network throughout East Asia, is still in the development stage. Because the project effectively bypasses the local PTTs, who have benefitted from a virtual monopoly in this area, it has encountered considerable local resistance. The project team had to seek support from senior government levels and, as a result of these efforts, government approvals of these efforts for the project were recently received.

In Japan, the proposed project was met by heavy opposition from Kokusai Denshin Denwa (KDD), the Japanese international telephone service provider. Having anticipated these difficulties, the company approached the Japanese government very early and stated its interest in a clear and candid manner. Though the KDD has strong government ties, its favored position has recently weakened. Consequently, the company's Japanese partner was able to exert substantially more influence, and the project received Japanese government approval. The Japanese partner has also been responsible for representing the project team in negotiations with the NTT and KDD.

The company views its success in the overseas market as the result of several factors. First, as a small business, the company chose to team up with other firms and take a project approach to the market. Second, the company stresses a flexible design approach and willingness to make any changes desired by the customer that are technologically practicable. Third, the company finds good local partners that can provide insight to local practices and help negotiate with local regulatory bodies. Lastly, the company always approaches the planning and costing of a project with the understanding that political and regulatory problems often result in costly delays. Thus, when delays do occur, the company is in a position to wait them out. It is the combination of these factors that has made the company's strategy a success.

NOTES

1. John F. Malone, "New Directions in Telephone Technology," Modern Office Technology, May 1985, pp. 57-72.

2. The Competitive Status of the U.S. Electronics Industry, prepared by the Electronics Panel, Committee on Technology and International Economic and Trade Issues of the Office of the

Foreign Secretary, National Academy of Engineering; and the Commission on Engineering and Technical Systems, National Research Council (Washington, D.C.: National Academy Press, 1984), pp. 79-80.

3. Ibid., p. 80.

4. Small Business Administration Data Base.

5. J. K. Paul, ed., High Technology International Trade and Competition (New Jersey: Noyes Publications, 1984), p. 115.

6. Ibid., pp. 118-120.

7. David E. Sanger, "Nippon Telegraph in Pact with IBM," New York Times, September 26, 1985.

8. Inside U.S. Trade, September 13, 1985, p. 12.

9. John Smalley, "Ringing Up New Profits," World (New York: Peat, Marwick, Mitchell & Company, July/August 1985), p. 19.

10. Paul, High Technology Trade, p. 122.

11. U.S., Department of Commerce, National Telecommunication and Information Administration, Telecommunication Policies in Seventeen Countries: Prospects for Future Competitive Access, No. NTIA-CR 83-24 (May 1983), pp. 101-114.

12. Smalley, "Ringing Up Profits," pp. 17-18.

13. Andrew Lloyd, "When the Planning Has to Stop," New Scientist, December 9, 1982, pp. 658-659.

14. Department of Commerce, Telecommunications in 17 Countries, pp. 203-213.

15. Kevin Smith, "Telecom Giants Edge Closer to Domination of World Markets," Electronics Week, May 27, 1985, pp. 38-39.

16. Department of Commerce, Telecommunications in 17 Countries, pp. 85-99.

17. Smalley, "Ringing Up Profits," pp. 21-22.

18. Inside U.S. Trade, September 6, 1985, p. 9.

19. John Gosch, "Europe Ready to Start R&D on Compatible Telecom Net," Electronics, August 19, 1985, p. 28.

20. Ibid.

21. Ibid., pp. 28-29.

9
OBSERVATIONS ON TRADE POLICY IMPLICATIONS

The research described here, as well as our previous NTB work for the National Science Foundation, and for the USTR on Japanese and European targeting in high-technology industries, has led us toward certain observations about the characteristics of trade barriers in technology-intensive industries, and the relationship between these barriers and the international behavior of U.S. firms. These are important as the government prepares for high-tech negotiations in the GATT, as well as ongoing bilateral discussions with the Japanese, the Europeans, the Brazilians, and others.

Nontariff barriers (NTBs) are important as barriers to the entry of major markets by U.S. high-tech companies, and significant in the calculations of these firms as they strategize around these barriers. Although growth asymmetries between major trading countries or misalignment of currencies are principally responsible for the size of the U.S. trade deficit, NTBs ought not to be removed from the negotiating agenda, since the long-term, negative competitive impacts of such barriers on U.S. companies --and small high-tech companies in particular--can be considerable. Nor should the fact that some companies adjust successfully to NTBs lead us to conclude that the "industry" knows best how to cope with these restrictions, since what constitutes success for a company in the short term may turn to failure over the longer term, while other firms may--through no fault of their own--fail to cope with NTBs even over the short term. We will discuss this in more detail later; it suffices here to say that there is no substitute for active U.S. government participation in the identification, analysis, and mitigation of nontariff barriers, and no possibility of this occurring without close cooperation between government and industry.

There are many justifications for responding to NTBs, of which two currently predominate within government. First, one may view such restrictions as incompatible with existing multilateral trade law or bilateral commitments, and seek various forms of redress as prescribed in treaty or in U.S. law. This legal approach has the advantage of precedent and international consensus (at least to some degree) but tends to be reactive, slow moving, and frequently ineffective for forms of trade restriction that lie outside the international legal regime. While this regime is dynamic and the subject of current U.S. efforts toward reform, it must of necessity lag trade practice, at least in those high-technology sectors in which market structure, competitive trends, and technology are changing rapidly.

Second, one may justify action based on the volume or source of industry complaints against foreign trade practice. This has been the case in mature industrial sectors and, increasingly so, in high-tech industries characterized by increasing import penetration (semiconductors in particular). The advantages that accrue here include a ready domestic constituency for action, and access to necessary data from the industry on the characteristics and impacts of the foreign practice in question. Unfortunately, action in response to such pressures is by definition reactive, may not always represent the interests of the industry as a whole, and may prove diplomatically embarrassing should the industry change its view in midstream. Most damaging, the dependence of action on industry complaint assumes that no news is good news; that the multitude of quiet, day-to-day adjustments that firms make in response to foreign trade barriers are in both the firms' and the countries' best interests; and that the judgments of firms (regardless of their level of experience, strategic acumen, financial strength, etc.) are wiser than those of government. The case studies described in this report deliver a very different message--specifically, the fallibility of corporate judgments and strategies, and the need for close interaction between firms and government agencies as trade barriers are confronted and overcome.

A third basis for action, one that we find more in keeping with the reality of NTB effects, and one more likely to produce consistent, long-term action, rests upon an improved understanding of the marketplace impacts of NTBs on individual firms and industries, both in the present and cumulatively over time. These barriers are subtle in their effects on corporate strategies, and go beyond the simple blocking of markets to influence corporate multinational strategies for investment, marketing, product development, technology diffusion, and long-term research. The effects of nontariff barriers are, therefore, likely to be manifested in ways not frequently expressed in public trade complaints by the industries, not easily convertible into balance-of-trade effects, and not particularly impressive quantitatively or qualitatively in the short term. The impacts that ought to concern us relate more to the long-term competitiveness of firms and sectors, the rate at which technology developed by U.S. firms diffuses to our competitors, market shares of U.S. firms in foreign markets over time, and the capacity of U.S. firms to get sufficiently close to foreign markets to anticipate competitive and regulatory developments, and to adjust to them in ways that protect their markets. It is these subtle, long-term competitive effects of NTBs that are least understood in both academic and government circles, but that provide the most legitimate long-term basis for government action.

What we have said about the subtleness of these effects, and the propensity of firms to make quiet adjustments over time to these distortions, suggests for negotiators an important principle of timing. There appears to be a point in the imposition of an NTB when individual firms (and in rare instances, an industry as a whole) are suffering negative effects and are prepared to lobby government for action; when data about the NTB are available; when the effects of these restrictions are concrete, visible, and, in some cases, quantifiable. This clearly is the optimum time for representations to the responsible government. In the absence of such action (or in cases in which the industry is insufficiently well organized to raise the issue in the first place), the restrictions induce adjustments by firms, which render the effects of the restriction progressively less visible. Some firms may opt out; others may license technology to local producers; some may joint venture with local partners; others may form wholly owned marketing or manufacturing subsidiaries. Firms employing these strategies to circumvent the NTB and access the market tend to lose interest in the barrier's removal, and some may develop an interest in its perpetuation. Government action becomes less and less likely as domestic pressure is relaxed, and access to information is restricted. The distortions created by the trade barrier continue to exist but become an integral part of the trading and investing relationship between the United States and the country involved. While at any given point these relationships may appear to be principally market driven, they are in many cases the consequence of previous foreign government actions designed to manipulate the conditions under which U.S. firms gain access to their market, in such a way that the domestic development objectives are satisfied. Thus, unless the window of negotiating opportunity is exploited at the time the NTB is imposed (or revealed by a tariff reduction, or develops as a consequence of administrative decision abroad), the opportunity may be gone, unless, at some future point, the United States chooses to develop new legislation that permits retaliation against previous trade practices with contemporary negative effects.[1]

Timing can be optimized in two ways: by anticipating the imposition of such barriers (which in turn requires greater understanding in government as to the sources of such restrictions); and by actively mobilizing industries and firms that are confronting such barriers daily in their foreign markets, but that, for various reasons, are unlikely to bring complaints about such distortions to the attention of government. Anticipation is rare, but not without precedent (the case of Japanese patent protection for software is a good example). For the most part, however, industry complains, government reacts. If it is the case that--as we have argued--NTBs, at least in high-technology sectors, are essentially the manifestations of industrial development priorities, then a fuller understanding of those priorities and the political and economic forces that shape them ought to permit anticipation of NTBs in those industries or technologies targeted by foreign governments.

This suggests an upgrading in skill and in our intelligence base with respect to foreign government industrial targeting strategies and objectives; it also suggests a change in orientation from viewing NTBs by barrier type (standards, procurement, licensing, and so on) to the monitoring of import policies and structures by sector. Such an orientation would acknowledge that governments establish product standards (or procure from local firms, or license imports, etc.) not because they have a weakness for these policy instruments, but because they are committed to building domestic productive capability in a certain sector. A sectoral orientation in the monitoring of foreign government trade policies would be more alert to new forms of protection, more concerned with actual trade performance in key sectors characterized by U.S. comparative advantage, and more likely to anticipate restrictions for those sectors/products/technologies targeted for domestic growth by our trading partners.

Improved anticipation would, in turn, permit the mobilization of private industrial constituencies that are politically dormant, either because they fail to recognize the negative impacts of NTBs or because they are badly organized. This presents a dilemma that the government has yet to confront. NTBs in high-technology industries can be extremely complex technically, can be subject to rapid change as the technology itself advances, can be the result of actions taken by obscure corners of a foreign government's bureaucracy--this in addition to the inevitable subtlety of NTB effects. These characteristics increase the need for real-time information from firms that are confronting these restrictions in their marketplaces, and require a level of government-industry cooperation that is usually present-- if at all--only immediately before and during major rounds of multilateral negotiations. It also, of course, increases the need for technical expertise in the major trade agencies, and/or close cooperation between the trade agencies and the technical agencies of government.

The dilemma here is that the firms themselves are not always aware of precisely what they are confronting in foreign markets, or if their awareness is high, may not be disposed to bring in the government if they believe the restriction can be circumvented through direct discussions with the foreign government involved. A clear message running through most of our cases is the inadequacy of corporate responses to foreign non-tariff barriers, even when judged in terms of the firm's own interests. Firms are frequently only dimly aware of the presence of a market distortion and tend to assume that the regulations they face are inherent parts of the marketplace rather than worthy subjects of complaint. Nor are the strategic adjustments that firms make to an NTB (through licensing, joint venturing, local manufacturing, and the like) necessarily rational or in the long-term interests of the firm; many of these adjustments appear to reflect inadequate understanding of the competitive dynamics of the market, of the nature of the barrier or distortion itself, or of the rights that the international trade regime confers on companies confronting such barriers. Firms attempting to deal on their own with such restrictions are frequently new to market, sometimes in financial circumstances that make them impatient to conclude agreements with foreign entities, and only rarely in a position to take advantage of the experience of other firms that have faced similar circumstances. Concerning this latter point, it was clear to us that few firms considering entry into the Japanese market either had access to or had made much effort to acquire information about what problems to expect, where to go for advice, what strategies might work in various circumstances, how to monitor success in the market over time, or how to deal with government agencies.

Thus, the dilemma is of two groups--government and business--each needing the other to respond effectively to NTBs, yet each laboring under many of the wrong assumptions, possessing inadequate information, each clinging to perceptions of the other that impede dialogue. As a result, there is substantial difficulty establishing and maintaining cooperation on these highly technical and politically charged issues.

Some of the answers were forthcoming in this research. The fact that NTBs are not perceived by many high-tech marketing officials as necessarily illegal or unusual suggests that government officials, in seeking information about NTBs, ought to couch their inquiries in broad terms. Our own experience in locating interview respondents suggests the same thing. Ask a businessman if he has confronted an NTB in a foreign market, and his likely response would be either "What's that?" or "Yes, U.S. export controls"; ask him if he has faced barriers to entering the market, and he will proceed to describe NTBs. The lesson here is not to be too legalistic in defining these barriers and to remember that the corporate sector is much more concerned with marketplace impacts of import regulation than with NTBs as defined by GATT codes of conduct and the like. The need in seeking such information is therefore to describe such barriers broadly with respect both to means of their implementation and their effects, and, with respect to the latter, to include not just market blocking effects but NTB-induced alterations in corporate strategies. Broadening the context in this way very likely would produce a richer and more significant NTB data base in government, a data base that would reflect the long-term strategic and competitiveness issues raised by foreign NTBs in high-tech industries.

A second implication of the way corporate officials define NTBs is that government agencies ought to actively seek such information from companies rather than assuming that the industry either will complain or deal with the barrier successfully on its own. To quote from our earlier work for the NSF, such an approach "ignores the small- and medium-sized firms, which seem disinclined to protest such restrictions in Washington, and frequently are incapable of circumventing restrictions abroad; and it assumes that distortions in corporate behavior associated with private adjustments to NTBs (licensing of technology, for example) involve no public, longer-term costs."[2]

The notion of more successfully engaging the cooperation of small- to medium-sized firms has consequences not only for the quality of NTB information but also for the degree of constituency support for free-trade negotiations. Again as noted in our earlier study, large firms familiar with the intricacies of foreign markets generally will not come to government for assistance in grappling with NTBs. Because of the large number of jobs, size of investment, and R&D resources that large firms can generate, they normally enjoy leverage in negotiating market access with foreign governments that small companies do not possess. They also command resources that permit them to sustain long periods of negotiation for market access, are of a size that enhances their bargaining position in discussions with potential joint-venture partners, and have a diversity of new technologies that makes the licensing of technology to foreign entities less threatening. If joint-venture arrangements and licensing possibilities do not work out, larger firms possess both the experience and the resources necessary to establish foreign manufacturing operations, a strategy that is available but frequently not seized by smaller firms. Thus, the options for market penetration available to large companies--short of involving the U.S. government--are numerous. The support that such firms can offer trade negotiators is therefore a product of complex corporate calculations and internal trade-offs, and may be lacking in both consistency and level of commitment.

Small firms, on the other hand, tend to have interests that are clear-cut; less subject to internal compromise among domestic and foreign manufacturing, product assembly, marketing and licensing operations; and more likely to view market-blocking barriers as problems without effective solutions, rather than as both problems and opportunities. The effective political mobilization of the small- to medium-sized business community, for access to information as well as for trade negotiating support, might, therefore, have very positive effects.

Finally, there are certain negotiating implications inherent in our findings concerning the nature and effects of NTBs. We have already discussed the issue of timing. It also bears emphasis that NTBs in high technology are manifestations of industrial policy objectives that define the core of international strategic and competitive priorities of foreign governments. Because such barriers reflect government commitments to national interests rather than domestic constituency pressures, they tend to be less tractable diplomatically, less amenable to trade-offs and compromise during multinational negotiations, and tend to resist pressures for implementation of broad free trade statements or codes of conduct. "Successful" negotiations on NTBs in high-technology sectors, therefore, tend to push governments toward more inventive means of protecting markets and stimulating domestic development; witness the stability of U.S. semiconductor market shares in Japan, despite successive trade negotiations and a highly competitive U.S. industry; witness also the consistency of French protectionism in high technology (from tariffs to quotas to licensing restrictions to standards and procurement practices) as the rationale for each form of protection was exhausted. Although it is very difficult to quantify a level of protection--particularly with respect to protection provided by NTBs--the qualitative evidence suggests that the actual marketplace is not significantly freer of such restrictions today than it was before the Tokyo Round negotiations.

If this is the case, one is led toward the controversial conclusion that market access is more appropriate as a negotiating principle than are agreements on specific forms of protection. Granted that the establishment of what constitutes legitimate market share, and how precisely governments might enforce bilateral or multilateral agreements on market shares, would be extremely complex, it remains that such agreements--negotiated sectorally--seem more consonant with events in the marketplace than do the current results of previous trade negotiations.

SOME ORGANIZATIONAL AND POLICY RECOMMENDATIONS

Recommendations will be discussed in four policy domains:

1. **Industry/Government Relations.** How does government acquire early warning, supporting data, case study material, and political reinforcement from the small business community? How can the government/industry relationship be placed on an ongoing, effective basis in the periods between major trade negotiations, when the mobilization of the business community becomes more problematic?
2. **Data and Analytical Resources.** What types of new data are required for timely government action on behalf of U.S. small firms confronting NTBs? How can we deal more effectively with the subtlety of NTB effects? What analytical approaches are necessary to capture the longer-term, competitive effects of trade barriers faced by small U.S. companies?
3. **Legal.** What gaps exist in the multilateral and U.S. legal framework that limit the remedies available to U.S. companies in attempting to deal with foreign NTBs?
4. **Diplomatic.** What ongoing diplomatic efforts deserve a greater emphasis? What new efforts might be made to eliminate or mitigate the effects of NTBs on small companies?

Industry/Government Relations. The problem in this area is that small- to medium-sized firms tend not to come to the trade agencies with complaints about nontariff barriers, at the same time that the agencies themselves are not particularly aggressive in seeking out this information. Small company complaints about discrimination in foreign markets therefore tend to be sporadic, disorganized, and unsustained. ISAC 14 has not been an effective mechanism for this particular purpose, though it may become more active in preparation for the GATT negotiations next year. In any event, the result of this lack of systematic communication is that opportunities for the mitigation of NTBs through diplomatic efforts are missed, and the effects of trade discrimination are built into the trade and investing relationship to such an extent that they become progressively less visible and therefore less amenable to action. We suggest the following:

- A greater government outreach program to access small- to medium-sized firms not represented on ISAC 14. This can be done through an expansion of ISAC membership, direct contact between government officials and individual firms, formal surveys of small- to medium-sized companies concerning their trade problems, and expanded contacts with other trade associations in which small companies are represented.
- The government should identify small- to medium-sized high-technology firms that have experienced discrimination in foreign markets, and are prepared to provide information on these episodes. These firms might form the basis of a panel to be periodically queried, or organized into a committee to periodically advise the trade agencies on NTB effects and necessary remedies.
- Techniques of eliciting information from firms must improve. We must recognize that "nontariff barrier" is not a term widely understood in business circles, nor is the typical firm likely to recognize the difference between impediments to market access that are legitimate and those that are not. Indeed, most firms we spoke to had a very dim understanding of their rights under international and domestic trade law. In eliciting relevant information from firms, it is therefore necessary to coach inquiries in general business terms (what are the problems you are facing in Market X? when did they start? what effect have they had? what have you done in response?) rather than in legal or diplomatic language.

Data and Analytical Resources. Our previous work for the NSF identified a number of gaps in the NTB data base relating to accuracy of information, internal consistency within data bases, currency of information, lack of quantification, and incompleteness. The current work confirmed these observations, and also found a serious lack of information in government concerning particular effects of NTBs on small- to medium-sized companies, which is a serious flaw particularly with respect to high-technology trade policy. We also found an analytical and conceptual problem, in the sense that lack of longitudinal data on particular trade barriers and a dearth of specific case studies prevented the government from understanding the long-term strategic, competitive effects of market protection. This latter problem is a very serious one, given that trade barriers imposed by foreign governments act with great subtlety, are both incremental and cumulative in their effects, and impinge on a wide range of corporate strategies not limited to exporting (including investing, licensing of technology, R&D decisions within the firm, and the like). An understanding of these effects is further complicated since these barriers may be applied for only a brief, but vital, period of time, during which the protected industry is able to take advantage of its guaranteed market to establish the basis for future competitiveness. In such instances, current trade complaints cannot reflect the current and future competitive effects of previously applied trade barriers. Furthermore, even when estimates of existing trade impacts are possible, they rarely capture the long-term, competitive effects of market closure, precisely the effects that ought to concern us in high-technology industries.

This suggests a number of areas in which the government's understanding of trade barriers and their effects must be deepened, through data collection, through studies, and through closer consultation with companies involved in foreign markets. They include the following:

- What are the current and prospective competitive effects of market protection in key high-technology industries?
- How has the form of protection evolved in these sectors (tariffs to quotas to government procurement, and so on), and are these different forms of protection designed for a single purpose, or do they emanate from different government objectives?
- How have these trade barriers influenced the behavior of U.S. companies, with an emphasis on small- to medium-sized firms?
- How can these effects be quantified?

This study has established that we need to understand more about the cost effects of various corporate adjustments to NTBs. These tend to include increasing the price of locally produced or marketed products, limiting the firm's proximity and therefore its sensitivity to local market conditions, blocking market access entirely, encouraging premature transfers of technology under agreements that do not protect the interests of the licensor, and distorting the allocation of R&D resources within the firms. These costs are likely to be more onerous for small companies.

The government should contribute to greater knowledge in these areas by:

- Sponsoring analytical studies in these areas, and/or encouraging other agencies, particularly the Commerce Department and the National Science Foundation, to sponsor such studies.
- Pressing for the improvement of existing NTB data bases at USTR and Commerce, in order to increase detail, quantification, historical information, and projections of potential competitive effects.
- Using the case studies in our study as a basis for a small business specific data base on NTBs, with cases updated periodically and new cases added as they are identified.

Legal Issues. This is not the occasion for a detailed evaluation of the adequacy of the multilateral legal framework for dealing with nontariff barriers. It suffices to say that the legal system as it relates to NTBs is in its formative stages, and that further specification and extension of GATT NTB codes, accompanied by streamlining of the adjudication process within the GATT and a buildup of case histories, will be necessary before NTBs are effectively constrained by multilateral agreements.

Of more relevance for our report are two areas in which neither the GATT framework nor U.S. trade legislation appears to be adequate. The first relates to those forms of home market protection--many of which are identified in our report--that are not dealt with effectively either multilaterally or through U.S. countervailing duty, anti-dumping or Section 301 statutes. A very interesting, and generally neglected report to the Congress by the U.S. Trade Representative (July 15, 1985) dealt with this issue in some detail. We quote from it below:

> First, the CVD law cannot be used effectively to offset foreign trade restrictions. It can deal with one form of home market protection--government procurement preferences in situations where the government pays more than the market price for the procured product. Such practices are not common, however, since as a general rule competitive bidding among domestic firms probably keeps the government from paying higher than market prices. Another potential limitation is that in

many cases the CVD law requires an injury test.[3] It is not likely that a U.S. industry would be able to demonstrate injury or threat thereof when preferential procurement policies are used to promote infant industries. Such industries are rarely in a position to export significant volumes of the U.S. market at the time the procurement occurs.

Second, the AD law does not address home market protection directly. However, an important consequence of home market protection is the possibility that domestic producers will attempt to price discriminate between domestic and foreign markets. To the extent that home market protection provides this additional commercial advantage, the AD law can be used to offset it. As in the case with the CVD law, the injury requirement may limit the effectiveness of the AD law in instances where the protected industry has not yet begun to export to the U.S. market in significant volumes.

Finally, Section 301 can be used only where the home market protection is deemed to be unfair under the law.[4] This excludes cases where the home market protection is "GATT-legal"; for example, high GATT-bound tariffs or preferential procurement by government agencies which do not fall within the entity coverage of Government Procurement Code are legitimate forms of home market protection under the GATT. It appears from the sectoral studies that some of the protection was permitted under the GATT and, therefore, was not unfair under Section 301.

It is beyond our scope to recommend new legal remedies in response to the imposition of foreign NTBs. We can say with confidence, however, that these barriers are onerous, particularly for small companies, that they have long-term competitive effects on our leading-edge industries, and that eliminating these restrictions through whatever approach would be of great assistance to small- to medium-sized firms.

The second area of some legal concern was touched on earlier, specifically the problem of reacting in the present to the effects of trade barriers imposed in the past. We have raised this problem as one of data gathering, since the competitive disadvantages that U.S. firms may be facing at any given point in time may appear to be the result of "free market forces," when in actual fact, foreign trade barriers no longer in clear evidence may be at play. The USTR report also addresses this issue as follows:

The second limitation in the trade laws is the inability to address targeting practices when they are separated from their market effects by time. The semiconductor study, for example, suggests that home market protection, R&D assistance, and joint R&D can enhance learning economies, thereby making recipient firms more competitive than they would be otherwise.

The economics works as follows: The home market is protected, thereby shielding domestic firms from import competition. In this less-risky environment, firms can reduce their variable costs through learning and scale economies in two ways:

1. They can increase production faster than they could in the face of competition.
2. They can achieve a higher volume of sales than they would if import competition were present.

Once firms are able to reduce variable costs by increasing the volume, they can become quite competitive in world markets and the protection can be eliminated. Hence, the effects of targeting policies may not be felt in international markets until long after the policies have been eliminated.

The CVD law may not be able to fully offset the delayed effects of past subsidy programs. When the Commerce Department allocates benefits from certain types of government programs over a number of years, its practice is to "front-load" such benefits. This allocation method may not coincide with the period in which the firms benefitting from the government programs begin to export to the U.S. market. For example, a $1.5 million grant for the purchase of capital equipment with a fifteen-year life would not be allocated over fifteen years in equal amounts (i.e., $100,000 per year). A higher allocation would be made in the early years when the equipment is more productive. Thus, if the foreign firm benefiting from the subsidy begins exporting and injuring U.S. firms in the fifteenth year of the equipment's life, a CVD would be imposed for only one year and the duty itself would be smaller than if it were imposed during the first year of the equipment's life. Any attempt to reallocate the benefits of government targeting to a later period, however, would almost certainly result in a GATT challenge. Moreover, allocating the subsidy solely over the period in which competitive benefits are deemed to occur would result in arbitrary rulings.

Section 301 may be used to deal with cases involving delayed effects within three limiting factors. First, the government practice must fall within the unfairness definition of Section 301. This would exclude "fair" practices such as GATT-legal home market protection. However, this limiting factor can be eliminated by amending the law to give the USTR negotiating authority such as that described in the previous section.

Second, the retaliatory authority of Section 301 could not be used to offset the delayed effects of targeting. The stated objective of Section 301 is ". . . to obtain the elimination of such (unfair) act, policy, or practice . . ." of a foreign government. Since those acts, policies, or practices that give rise to the delayed effects often no longer exist, the President could not retaliate under Section 301. Moreover, such retaliation would undoubtedly be challenged in the GATT.

Finally, it is not clear what kind of counteraction is economically optimal to offset delayed effects. At this time, there is no generally accepted methodology in the field of economics for quantifying these delayed effects. More research, some of which is being considered in academia, needs to be done on the economic impact of various targeting practices to resolve this question. Hence, amending the law to deal with delayed effects is premature at this time.

Again, we are in no position to comment on legislative remedies for this problem of delayed effects. We would only emphasize that it is a serious issue for small- to medium-sized firms, particularly in high-tech industries.

Diplomatic Approaches. NTBs in high-technology industries have been at the center of recent bilateral negotiations with the Japanese in particular, and an important aspect of preparations for the major round of multilateral negotiations more likely to take place later this year. The bilateral discussions--dubbed the MOSS talks or market-oriented, sector-specific--have been useful in identifying specific types of nontariff barriers of concern to

U.S. high-tech companies in Japan, and in achieving some liberal-ization, particularly in standards and procurement. Intellectual property protection is another area that should now receive greater focus in the second round of negotiations, very likely to commence in the near future. Whether Japanese concessions made in these negotiations will have a concrete effect on market access remains to be seen. This uncertainty further increases the importance of government/industry contacts as the implementation process begins.

NOTES

1. The Office of the United States Trade Representative, "Report on Foreign Targeting," unpublished report to Congress, dated July 15, 1985.

2. Robert Cohen, Richard Ferguson, Michael Oppenheimer, Nontariff Barriers to High-Technology Trade (Boulder: Westview Press, 1985), p. 5.

3. In CVD cases, the injury requirement applies only to countries that have signed the Subsidies Code or that have assumed obligations with the United States that are substantially equivalent to obligations under the Subsidies Code. All countries receive the injury test under the AD law. These conditions apply throughout this report to the CVD and AD laws.

4. Unfair home market protection under Section 301 includes protection which ". . . denies benefits to the United States under any trade agreement, or is unjustifiable, unreason-able, or discriminatory and burdens or restricts U.S. commerce." If the protection meets this standard, the President can retaliate if the foreign government involved fails to eliminate the offensive practice.

ABBREVIATIONS

AD	Antidumping
AT&T	American Telephone and Telegraph
BABT	British Approvals Board for Telecommunication
BMFT	Bundesministerium fur Forschung und Technologie
BSI	British Standards Institution
BT	British Telecom
BTG	British Technology Group
CAD/CAM	Computer-aided design/computer-aided manufacturing
CCITT	Comite Consultatif International Telephonique et Telegraphique
CEN	European Committee for Standardization
CEPT	Conference Europeene des Postes et Telecommunications
CGE	Compagnie Generale d'Electricite
CGS	Cap Gemini Sogeti
CII	Compagnie Internationale pour l'Informatique
CMOS	Complementary metal-oxide semiconductor
CVD	Countervailing duties
DBP	Deutsche Bundespost
DEC	Digital Equipment Corporation
DGT	Direction Generale des Telecommunications
DTI	Department of Trade and Industry
EDP	Electronic data processing
EEC	European Economic Community
EMI	Electromagnetic interference
EPC	European Patent Convention
ESPRIT	European Strategic Program for Research and Development in Information Technologies
GATT	General Agreement on Tariffs and Trade
GEC	General Electric Company

GOF	Government of France
GOJ	Government of Japan
HB	Honeywell-Bull
IBC	Integrated broadband communications
IBM	International Business Machines
IC	Integrated circuit
ICL	International Computer Ltd.
INS	Information Network System
IPA	Information Technology Promotion Agency
IRIA	Institut de Recherche d'Informatique et d'Automatique
ISAC14	Industry Sectoral Advisory Committee (Small and minority business group)
ISDN	Integrated services digital networks
ISO	International Standards Organization
ITT	International Telephone and Telegraph
JDB	Japanese Development Bank
JECC	Japan Electronic Computer Company
LAN	Local area network
MAP	Microelectronics Applications Project
MISP	Microelectronics Industry Support Program
MITI	(Japanese) Ministry of International Trade and Industry
MOS	Metal-oxide semiconductor
MOSS	Market-oriented, sector specific
MTN	Multilateral trade negotiation
NBF	New biotechnology firm
NCR	National Cash Register
NEC	Nippon Electric Company
NMOS	Negative metal-oxide semiconductor
NTB	Nontariff barrier
NTT	Nippon Telephone and Telegraph
OEM	Original equipment manufacturer
OSTP	Office of Science and Technology Policy
OTA	Office of Technology Assessment
PTT	Ministry of Postes and Telecommunications
RACE	Research and Development in Advanced Communications Technologies for Europe
R&D	Research and development
RFP	Request for proposal
SBA	Small Business Administration
SEL	Standard Electric Lorenz
SEMI	Semiconductor Equipment and Materials Institute
SERC	Science and Engineering Research Council
SG2	Societe Generale de Service et de Gestion

SPERAC	Systemes et Peripheriques Associes aux Calculateurs
STA	Science and Technology Agency (Japan)
STC	Standard Telephone and Cables
STK	Standard Telephone Kablefabrick
TDMA	Time division multiple access
TI	Texas Instruments
USTR	U.S. Trade Representative
VANs	Value-added networks
VDE	German Society of Engineers
VLSI	Very large scale integrated

BIBLIOGRAPHY

"Apple Computers vs. French Chauvinism: Politics, Not Free Trade, Wins in the End." Wall Street Journal, March 1984.

Baldwin, Robert E. Non-Tariff Distortions of International Trade. Brookings Institution, 1970.

Bio/Technology, April 1984, p. 33.

Cane, Alan. "U.S. Imports Dominate the Market." Financial Times, June 24, 1985, p. 6.

Choy, Jon. "Tokyo Expands Copyright Law to Cover Software." JEI Report No. 28 B, July 19, 1985.

Cohen, Robert; Ferguson, Richard; and Oppenheimer, Michael. Nontariff Barriers to High-Technology Trade. Boulder: Westview Press, 1985.

The Competitive Status of the U.S. Electronics Industry. National Research Council. Washington, D.C.: National Academy Press, 1984.

Electronics Industry Association. "Changes in U.S. Telecommunications Industry and Impact on U.S. Telecommunications Trade." Submitted to the International Trade Commission, April 17, 1984.

"Electronics Newsletter." Electronics, August 19, 1985, p. 12.

Farnoux, Abel. Excerpts from the Report by the Committee for Electronics. Paris: French Ministries of State and of Research and Technology, March 1982.

"French Ready Translation System." Electronics, July 1, 1985, p. 27.

The Futures Group. "Data Base on Industrial Targeting." Glastonbury, Conn.: The Futures Group, 1984.

_____. The Impact of Foreign Industrial Targeting on the U.S. Computer Industry, Report 701-138-03. Glastonbury, Conn.: The Futures Group, April 30, 1985.

Gosch, John. "Europe Ready to Start R&D on Compatible Telecom Net." Electronics, August 19, 1985, p. 28.

Grey, Rodney de C. "A Note on U.S. Trade Practices." In Trade Policy in the 1980s, edited by William R. Cline. Washington: Institute for International Economics, 1983.

Howard, James S. "Deep Pockets." D&B Reports, July/August 1984, p. 24.

Inside U.S. Trade, September 13, 1985, p. 12, and September 6, 1985, p. 9.

The Labor-Industry Coalition for International Trade. Performance Requirements. Washington, D.C.: LICIT, March 1981.

Lloyd, Andrew. "When the Planning Has to Stop." New Scientist, December 9, 1982, pp. 658-659.

Logica. "Crowning Europe's DP Royalty." Datamation, September 1, 1984, pp. 157-160.

Malone, John F. "New Directions in Telephone Technology." Modern Office Technology, May 1985.

Organization for Economic Cooperation and Development. Committee for Information, Computer and Communications Policy. "Software: A New Industry" (draft). Paris: OECD, February 2, 1984.

Organization for Economic Cooperation and Development. The Semiconductor Industry, Trade Related Issues. Paris, 1985.

Paul, J. K., ed. High Technology International Trade and Competition. New Jersey: Noyes Publications, 1984.

Sanger, David E. "Biotechnology's Patent War." New York Times, March 19, 1984, pp. D1, D5.

_____. "Nippon Telegraph in Pact with IBM." New York Times, September 26, 1985.

Small Business Administration Data Base.

Smalley, John. "Ringing Up New Profits." World. New York: Peat, Marwick, Mitchell & Company, July/August 1985.

Smith, Kevin. "Britain Makes Major Bid to Build Commercial Fifth-Generation Machine." Electronics, July 8, 1985, pp. 26-27.

_____. "Telecom Giants Edge Closer to Domination of World Markets." Electronics Week, May 27, 1985, pp. 38-39.

Trewhitt, Jeff. "U.S. Midwest Makes Pitch for Biotech-Research Firms." Chemical Engineering, April 15, 1985, pp. 27-28.

United Nations. Center on Transnational Corporations. Transnational Corporations in the International Semiconductor Industry, 1984.

U.S. Congress. Office of Technology Assessment. Commercial Biotechnology: An International Analysis, OTA-BA-218. Washington, D.C.: Government Printing Office, January 1984.

U.S. Department of Commerce. Bureau of Industrial Economics. French Government Assistance to Its Domestic Computer

Industry: Lessons and Implications, BIE/IR-83-01, June 1983.

U.S. Department of Commerce. CMP Industry Sector Analysis, by Myles Denney-Brown. Washington, D.C.: Government Printing Office, August 6, 1984.

U.S. Department of Commerce. International Trade Administration. A Competitive Assessment of the U.S. Software Industry, December 1984.

_____. High Technology Industries: Profiles and Outlooks, Biotechnology, Table 11 (draft), July 1984.

_____. International Market Research Profile on Mini and Micro Computer Systems in France, March 1984.

U.S. Department of Commerce, International Trade Commission. International Developments in Biotechnology and Their Possible Impact on Certain Sectors of the U.S. Chemical Industry. USITC Pubn. No. 1589, October 1984.

U.S. Department of Commerce. National Telecommunication and Information Administration. Telecommunication Policies in Seventeen Countries: Prospects for Future Competitive Access, No. NTIA-CR 83-24, May 1983.

U.S.-Japan Trade Study Group. "Progress Report: 1984," September 1984.

U.S. Trade Representative. Interviews with officials, April/May 1984.

_____. Japanese Trade Barriers to U.S. Trade and Recent Japanese Government Trade Initiatives, November 1982.

_____. "Report on Foreign Targeting." Unpublished report to Congress, July 15, 1985.

Uttal, Bro. "Japan's Latest Assault on Chipmaking." Fortune, September 3, 1984, pp. 78-80.

Walton, Kim Alayne. "Cloning Around, Dr. Panglass Is Alive and Well in Bio-Tech." Barrons, January 7, 1985, p. 14.

Wiegner, Kathleen. "Europe Fights Back." Forbes, August 12, 1985, p. 82.

Williamowski, Karen. "CMP Industry Sector Analysis Form," July 17, 1984, pp. 1-7.

Ypsilanti, Dimitri. "The Semiconductor Industry." OECD Observer. OECD, January 1985.

ABOUT THE AUTHORS

Michael F. Oppenheimer is Executive Vice President at The Futures Group, with principal responsibilities for international marketing, project management, and corporate policy. Mr. Oppenheimer has managed projects on North-South negotiations, U.S. technology transfer policy, future strategic issues, and U.S. trade policy with respect to export credit subsidies and high-technology trade. He has consulted for the Departments of Commerce and State, the U.S. Trade Representative, the Office of Science and Technology Policy, and the Office of Technology Assessment, as well as numerous private firms. He is co-author of books on the U.S. Export-Import Bank, U.S. diplomacy in global conferences, and nontariff barriers to high-technology trade. Prior to joining The Futures Group, he was international relations advisor to the General Accounting Office and did graduate work in foreign policy at the University of Virginia.

Donna M. Tuths is a Senior Research Associate in the International and Technology Studies Section of The Futures Group. Ms. Tuths conducts research for both corporate and government clients on a wide range of topics including comparative technology assessment, foreign industrial targeting measures, worldwide market shares in select high-technology sectors, and international strategic business planning. Ms. Tuths earned a Master's degree from the Johns Hopkins School of Advanced International Studies, where she studied at the Center for European Studies in Bologna, Italy, and an AB from Barnard College. Prior to joining The Futures Group, she worked in Strategic Planning at Sikorsky Aircraft.